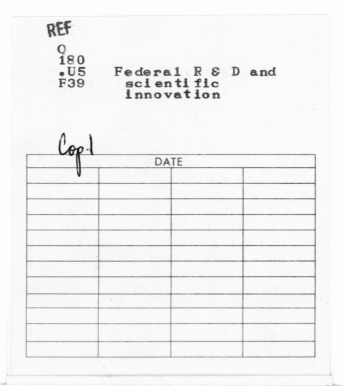

Federal R&D and Scientific Innovation

Leonard A. Ault, EDITOR
NASA

W. Novis Smith, EDITOR
Air Products and Chemicals, Inc.

Based on a symposium
sponsored by the Division
of Industrial and Engineering
Chemistry at the
176th Meeting of the
American Chemical Society,
Miami Beach, Florida,
September 11–14, 1978.

ACS SYMPOSIUM SERIES **105**

AMERICAN CHEMICAL SOCIETY
WASHINGTON, D. C.   1979

Library of Congress CIP Data

Federal R & D and scientific innovation.
  (ACS symposium series; 105 ISSN 0097–6156)

  "Based on a symposium sponsored by the Division
of Industrial and Engineering Chemistry at the 176th
meeting of the American Chemical Society, Miami
Beach, Florida, September 11–14, 1978."
  Includes bibliographies and index.

  1. Federal aid to research—United States. 2. Techno-
logical innovations—United States.
  I. Ault, Leonard A., 1933–   . II. Smith, William
Novis, 1937–   . III. American Chemical Society.
Division of Industrial and Engineering Chemistry. IV.
Series: American Chemical Society. ACS symposium
series; 105.

Q180.U5F39   338.4'7'607273   79–13114
ISBN 0–8412–0507–8   ACSMC8 105 1–184 1979

PRINTED IN THE UNITED STATES OF AMERICA

ACS Symposium Series

Robert F. Gould, *Editor*

FOREWORD

The ACS SYMPOSIUM SERIES was founded in 1974 to provide a medium for publishing symposia quickly in book form. The format of the Series parallels that of the continuing ADVANCES IN CHEMISTRY SERIES except that in order to save time the papers are not typeset but are reproduced as they are submitted by the authors in camera-ready form. Papers are reviewed under the supervision of the Editors with the assistance of the Series Advisory Board and are selected to maintain the integrity of the symposia; however, verbatim reproductions of previously published papers are not accepted. Both reviews and reports of research are acceptable since symposia may embrace both types of presentation.

CONTENTS

PREFACE

Over 50% of all research and development funds in the United States originates from the federal government. How to most efficiently effect commercialization and utilization (innovation) of this large amount of research and development remains an ongoing challenge.

The unique nature of federally funded R&D, in addition to its size, makes this a special topic in its own right. A number of problems for commercialization and, therefore, innovation are similar to industrially supported research. However, a great many more problems are not related, including ownership of patent rights, goal-oriented programs not related to the commercial market, lack of incentives, questions as to the extent of government involvement, etc.

In order to obtain current thinking, experience, and comments relating to the commercialization of federally funded R&D, the Division of Industrial and Engineering Chemistry of the American Chemical Society sponsored a symposium on this topic "The Commercialization of Federally Funded R&D" during the National meeting held in Miami Beach, Florida.

The participants in this symposium were selected because they represented either government or nongovernment organizations, and because they were involved directly in the problem of commercialization of federally funded R&D.

Because of outside interest in this symposium, we have added related papers for the publication of this book. "Innovation" has been added to the title to better reflect the relationship of these papers to what will be the long standing concern of the entire research and development area of the United States—government and private sector.

This book is not the final word, but is an initial statement by many of the participants who are directly involved in and concerned with ways to more extensively and effectively utilize the results of federally funded R&D.

LEONARD A. AULT
NASA
600 Independence Ave., SW
Washington, DC 20546
March 13, 1979

W. NOVIS SMITH
Air Products and Chemicals, Inc.
Box 538
Allentown, PA 18105

CONTRIBUTING AUTHORS

LEONARD A. AULT is Chief, Dissemination and Analysis Branch, Technology Transfer Division of NASA. He has been actively involved in the problems of technology utilization and innovation since 1963 shortly following the initial inception of the NASA Technology Utilization Program. Prior to joining NASA, he was a research physicist for the Harry Diamond Laboratories, U.S. Army Ordnance.

ALDEN S. BEAN is Director of the Division of Policy Research and Analysis at the National Science Foundation. Before joining NSF in 1973, he taught at the State University of New York at Albany and at Northwestern. He holds PhD and MS degrees in Management and Industrial Engineering from Northwestern University, and he has a BA in Economics from Lake Forest College.

WILLIAM D. CAREY is Executive Officer of the American Association for the Advancement of Science and Publisher of SCIENCE Magazine. Previously, he was a vice-president of Arthur D. Little, Inc., following a long career in the Bureau of the Budget where he was Assistant Director with general responsibilities for federal science policy.

ROBERT J. CREAGAN, Director for Technology Assessment, has made significant contributions working in many R&D projects at Westinghouse Electric Corporation, including the first nuclear submarine (Nautilus), the first commercial nuclear power plant (Yankee), fusion, and superconducting generators.

WILLIAM M. DOANE received his PhD in Biochemistry from Purdue University in 1962. Since that time he has conducted and led research in isolation, characterization, and modification of natural polymers, principally starch. Currently he is Research Leader of Derivatives and Polymer Exploration Research at USDA's Northern Regional Research Center.

DOUGLAS HARVEY is Director of Industrial Programs under the Office of the Assistant Secretary of Conservation and Solar Applications, Department of Energy. He has been involved directly in the management of the DOE/ERDA programs concerned with the commercialization of industrial conservation technology for over four years.

ARLEN J. LARGE has been a member of the Wall Street Journal's Washington bureau for 20 years, at various times covering the Departments of Agriculture, Treasury, and Commerce, plus Congress and many political campaigns. Currently he specializes in the relationship between science and the government.

ALBERT MAASBERG graduated from the New York State College of Forestry at Syracuse University with a BS degree and immediately joined the Dow Chemical Company at Midland, MI. He has worked there continuously in research, development, production, and administration. For the past 15 years he has been Director of Contract Research, Development, and Engineering for Dow.

CLYDE McKINLEY has been with Air Products and Chemicals in Research and Development roles for over 25 years. Much of this period has been in cryogenic technology, basic properties studies, and process development. His personal involvement in the liquid hydrogen program in America provides a unique perspective for his chapter.

STEPHEN A. MERRILL works on science and technology issues for the Senate Committee on Commerce, Science, and Transportation. A political scientist, he is a graduate of Columbia, Oxford, and Yale Universities. Formerly, he was a Congressional Fellow of the American Political Science Association, a Brookings Institution Research Fellow, and a staff member of the Senate Select Committee to Study the Committee System.

MICHAEL MICHAELIS is a member of the Senior Staff of Arthur D. Little, Inc. His work is primarily concentrated at the interface between government and industry: i.e., on governmental planning, programming, and decision making. He also is involved in the formulation of public and private policies and action programs to achieve national and corporate goals through fullest use of technical, institutional, and social innovation. He was Executive Director of the White House Panel on Civilian Technology, 1961–1963, and Executive Director of the Research Management Advisory Panel, Committee on Science and Technology, U.S. House of Representatives, from 1963 until the present.

CHARLES F. MILLER is the Technology Utilization Officer for the Department of Energy's Lawrence Livermore Laboratory, Livermore, CA. He also serves as the Coordinator, Far West Region, Federal Laboratory Consortium for Technology Transfer. He attended the University of California, Los Angeles, where he received BS and MBA degrees in Production Management. He has been with the Lawrence Laboratories at Berkeley and Livermore with assignments in Data Processing Systems, Mechanical Engineering, Theoretical Physics, and most recently as Program Manager for Technology Applications. Mr. Miller has published formal reports, progress reports, and proposals in areas such as Technology Transfer Methodologies, Air Pollution Modeling, Environmental Studies, and Identification of Criminal Explosives.

RICHARD PENN is Director, Center for Field Methods, National Bureau of Standards. Prior to this, he was Chief of the Decision Methods Group of NBS and was involved in operations research regarding science and technology.

WILLIAM O. QUESENBERRY was Assistant Chief of Naval Research for Patents and Patent Counsel for the Navy until he retired in December 1978. He joined the Office of Naval Research from private practice of patent law in 1951. Since that time, except for a period during 1968–1969 when he served as the first Director of the Office of International Patent and Trademark Affairs for the Department of Commerce, he has been with the Navy Patent Organization of ONR in various field and headquarters positions. He served during World War II with the Army Corps of Engineers and later in a reserve status with the Army Judge Advocate General's Corps. Mr. Quesenberry received his Engineering degree from Virginia Polytechnic Institute and his Juris Doctor and Master of Laws degrees from the George Washington University National Law Center.

J. DAVID ROESSNER is Senior Policy Analyst in the Policy Analysis Branch of the Solar Energy Research Institute. From 1973 through 1978 he was a Policy Analyst in the Division of Policy Research and Analysis at the National Science Foundation. He holds MA and PhD degrees from Case Western Reserve. He also has an MS in Electrical Engineering from Standard and a BS in Electrical Engineering from Brown University.

W. NOVIS SMITH is Assistant Director of Contract Research for Air Products and Chemicals, Inc. in Allentown, PA. Dr. Smith has been involved in all aspects of industrial research relating to the chemical industry for approximately 20 years. He holds 23 U.S. patents and has published a number of technical articles.

GEORGE TOLLEY is a professor in the Department of Economics and Cochairman of the Resource Analysis Group at the University of Chicago. He is the author of numerous articles and books on benefit–cost analysis. During 1974 and 1975 he was the Deputy Assistant Secretary of Treasury for tax policy.

STUART TOWNSEND is a graduate student at the University of Chicago and an economist for WAPORA, Inc.

INTRODUCTION

Innovation has entered the language of national politics, and it is a term that begs to be understood. A baffling set of dilemmas involving the relationships between the market economy and government surround the choices to be made in the sphere of public policy. Straight-line solutions are suspect because innovation involves as much art as invention, and because a multitude of institutional forces converge on the process of innovation. Among them are the legal system, economics, social policy, management, and politics.

We have been here before. It is not the first time that the question of the role of government in influencing the shape and quality of the industrial economy has been before us. Too often it has been the case that governmental intervention has been of the adversarial kind. Now we are observing the discovery by government that innovation suffers from some kind of drag, and the problem is to distinguish between government-induced causation and that which arises from within the industry sector itself. It will not be easy, and it may not be done quickly. It remains to be seen what innovation needs most: public policy action or public policy reform.

There is a degree of consensus at the core of the debate. In terms, it admits to a shared apprehension that the historical dynamics of industrial risk-taking, new market formation, and technological innovation are not working according to form, and that the resulting decline in innovative vitality spells bad news for the future worth and advancement of the national economy. Surface signs of a genuinely ailing economy are plainly visible in the tortured state of the dollar on the international exchanges, dismal productivity, and tenacious inflation. Coupling this syndrome with anxiety over innovativeness and a prevailing business climate that hedges risk-taking may be, on the one hand, a case of mixing chalk and cheese or, on the other hand, an admirable flash of intuition. It is very hard indeed to dismiss the probability of a connection.

Whatever may ail the once rampant dynamic of U.S. technological exuberance, and whatever the superficial or fundamental remedies, astonishingly little mind is being paid, in high echelons of economic policy management, to the function performed by research, development, and innovation in influencing the performance, near or long term, of the national economy. Though the point has been taken at the political level in President Carter's summons to "a new surge of technological innovation," it has not shown up conspicuously in the essays of his economic general staff. The field of policy attention is limited, on government's side, to the Commerce Department, the President's Science Adviser, the National Science Foundation, and scattered interest in the Congress. As for the business sector, there has been no dearth of alarm and less reluctance to indict flawed public policy as the source of the mischief.

The old myth about the separateness between the "private" and "public" sector was demolished long ago. The U.S. market economy is far from resembling the classic free market. Its performance is heavily socialized and politicized, both directly and indirectly through government's influence on the climate of risk and benefit, to say nothing of the play of such externalities as foreign energy pricing and supply. All this, coupled with the sophistication of decision analysis systems in corporate resource allocation, sharpens the sensitivity of business to the uncertainties and contradictions of the public sector. Though the environments and the working premises of the two sectors are poles apart, they mingle and traffic in the real world in a way that suggests nothing as much as the scientific phenomenon known as the Brownian movement.

Research and development strategies of government and industry might, in a rational political economy, be complementary. In the case of major competitors and adversaries of the U.S., they are indeed; but not here. Whether this is good or bad, for us is a debaters' argument laced with opposing premises. To bring proprietary R&D within some orbit of combined public/private rationalization might simply start us on a long journey to nowhere. Conversely, the total lack of combined strategy may lie precisely at the heart of the disruption of innovative capacity and a drifting national economy.

An introduction is no place to settle that argument. The provocative papers which the American Chemical Society has assembled from its 1978 Symposium on "The Commercialization of Federally Funded R&D" serve better to draw the lines and examine the predicament from a wide spectrum of thought, evidence, and opinion. Though the topic is centered on the role of federally funded R&D in generating commercialization, the authors have not been shy in addressing the larger context of problems of choice in rationalizing the infrastructure of innovation. All sides are heard from: industry, government agencies, Congressional staff, and independent experts whose qualifications are more than ample to contribute to the discussion. One can hope that our harassed policy makers in board rooms, in the Administration, and in the Congress will have the interest and the open minds to reflect on what is here.

American Association for the Advancement   WILLIAM D. CAREY
of Science
Washington, DC 20036
March 13, 1979

OVERVIEW

The Political Nature of Civilian R&D Management[1]

STEPHEN A. MERRILL

Committee on Commerce, Science, and Transportation, United States Senate, Washington, DC 20510

Contributors to this symposium and other discussions and studies preceding it share a belief in the desirability of commercializing the products of Federal research and development efforts as a way of serving important public needs and increasing the productivity of Federal expenditures. They are concerned that the results so far are mixed; few doubt that efforts to transfer Federal R&D products to the private sector have encountered difficulties and fallen short of their potential. Often the conclusion is that we must systematically identify the barriers to commercialization, whether in government policies and program management or in the market, and devise ways of overcoming them. It is presumed that program and project managers will follow effective innovation strategies if they are made aware of them. The implication of these assumptions is that the issue is one of means, not ends.

A number of observations suggests otherwise, at least with regard to that part of the Federal R&D effort whose purpose is to produce widely distributed social benefits, primarily through the commercialization of new products, processes and services. The growing criticism of direct government interventions in the

---

[1] The views expressed in this paper are those of the author, but they reflect the broader concerns of the Senate Commerce, Science, and Transportation Committee and, in particular, its Subcommittee on Science, Technology, and Space. The Subcommittee was reconstituted in 1977 as a result of the sweeping Senate reorganization, which enlarged the jurisdiction of the Commerce Committee by giving it legislative authority and oversight responsibility for NASA and Federal research and development policy generally as well as the Office of Science and Technology Policy and the science and technology activities of the Commerce Department.

market is not confined to regulation but extends to presumably
supportive activities, including civilian applied research and
development. The President's FY 1979 budget message notwithstanding, the Administration has yet to take an entirely consistent position on the role of publicly funded R&D in stimulating
innovation in the civil sector. Congress, too, is ambivalent.

This paper does not question previous government decisions
to invest heavily in civilian R&D nor suggest what future spending priorities should be; but its premise is that commitments
have often been made without agreement on goals and means and
consequently with disappointing results. Increasing the desired
returns on future expenditures requires the development of criteria for effective government intervention and appropriate
arrangements for government-industry collaboration. The emerging
national concern about the contribution of industrial innovation
to America's economic growth and international trade position
creates an unusual opportunity to design better procedures for
planning and executing the broad range of civilian R&D programs.

Progress in this direction depends, however, upon a clear
recognition of the public and private interests at stake. In
degree, both must be served if an R&D project is to result in a
socially beneficial innovation; but inevitably, the values and
methods of operation of government and private firms are partly
at odds. On the government side, these are matters of law, regulation and organization as well as professionalism and tradition.
Since the essence of the manager's job -- to exploit common
interests, accommodate differences, and resolve conflicts -- is
political in the broadest sense, it follows that the most important problems of R&D management are, inescapably, basic issues
of public policy.

The Context of Civilian Research and Development

More than two-thirds of Federal R&D spending is for research
and technology of use to the government in defense, space and
other national missions. Undoubtedly, much of this work has
found important civilian applications, for example, in aviation,
nuclear energy, and electronics. (1) In an accompanying paper,
Rubin Feldman describes the formation of a new firm based on the
application of certain coatings designed for use in space flight
to the protection of ordinary construction materials from fire.
Much of the technology developed for government use, nevertheless,
has not found eager takers in the private sector. NASA's Technology Utilization Program and a variety of other smaller-scale
efforts are attempting to promote such "spin-offs."

No one, least of all in Congress, disputes the desirability
of transferring technology to the private sector; but it is
equally clear that Federal spending on R&D for the government's
use must be justified as meeting the government's often esoteric
requirements and doing so efficiently. Not only is the govern-

ment the purchaser and user of this R&D, in many cases it is also
the performer. Alternatively, government employs the resources
of the private sector through elaborate procurement systems, some
of them designed to suit the needs of particular agencies and
programs. The procurement process has been modified to serve
other social goals, such as equal employment opportunity; but
commercialization is not one of them. In view of these constraints, it should come as no surprise that private industry has
not exploited a great deal of this technology.

As for informal and formal technolgoy transfer efforts, it is
difficult to judge their effectiveness. One can try to compare
their direct costs with resulting commercial sales and corporate
tax revenues, but what ratio is achievable and what size of program is optimal? In the absence of controlled experiments or
estimates of the stock and potential value of government-acquired
technology, we simply do not know. Thus, while Congress is dis-
posed to support technology transfer activities, its judgments
are unfortunately, but perhaps inevitably, guesswork.

Over the past decade, in any event, military and space R&D
spending have dropped significantly in constant dollars and in
relation to the growing "civilian" portion of the Federal R&D
budget. Government spending on health, energy, transportation,
housing, agricultural, environmental and other civilian R&D
projects now exceeds $9 billion annually. It has grown from 23%
of the R&D budget in FY 1969 to an estimated 38% in FY 1979. (2)
With the exception of some basic research, this public investment
is intended to meet pressing public needs, primarily through innovations in the private sector. Often commercialization is essential even where the products of the R&D are to be used largely or
exclusively by Federal, State or local governments, for example,
in mass transportation, education, and law enforcement.

There is a tendency to think of government intervention in
the civil sector as a response to social crises or market failures, as in the case of energy; but in nuclear power, agriculture,
health and other areas, the government has undertaken major R&D
projects largely because of perceived new opportunities, even if
these investments are justified in part by the inability or unwil-
lingness of the private sector to finance them. Whichever the
rationale, the civilian R&D budget will very likely continue to
grow and become more diversified. With the advent of the space
shuttle system, for example, the U.S. is on the verge of a variety
of space applications in which the private sector will play important roles. These include, in all probability, greatly expanded
global information and communications systems and, conceivably,
solar energy transmission and even space manufacturing. (3)

The Role of Government

The growth of the civilian R&D budget does not signify a consensus regarding the proper extent of government intervention in

the civil sector or the role of Federal R&D in particular. Consider, for example, the present Administration's views as reflected in the fiscal year 1979 budget and other recent initiatives. In January 1978, President Carter proposed a significant increase in basic research funding while restraining applied research and cutting back development projects and civilian R&D in real terms. The recommended increases were 10.9%, 7.4%, 4.6%, and 2.4%, respectively, compared with a then expected inflation rate of 6%. The Administration defended the slowdown in applied and civilian R&D spending on the basis of "the need to avoid overtaking activities that are more appropriately those of the private sector, such as developing, producing, and marketing new products and processes . . .," as well as the need to avoid duplication and failure. (2)

What was odd was the justification of the increase in basic research as a stimulus to innovation. In his State of the Union message, the President made the encouragement of "a new surge of technological innovation by American industry" the goal of his recommendations for "a program of real growth of scientific research and other steps to strengthen the Nation's research centers . . ."; and by implication, at least, the budget documents reiterated this message. (4) Basic research spending can be justified on various grounds; yet as the Director of the Congressional Office of Technology Assessment, Russell Peterson, recently pointed out, the economic argument is weak. The effects of advances in knowledge are usually difficult to trace and, for the most part, long term. (5)

If the pace of innovation does threaten our economic welfare, why did the President not recommend a comparable boost for applied civilian research and development which, in theory, have a much more direct bearing on commercialization? Perhaps the answer is that the Administration prefers to leave development largely to the private sector, ameliorate some of the negative effects on innovation of current Federal regulatory, economic, tax, and antitrust policies, and institute indirect incentives by way of creating a more favorable economic climate. This is one possible outcome of the mammoth interagency innovation policy study, which the President launched in May 1978 and directed to produce recommendations by April 1, 1979.

Simultaneously, however, the Assistant Secretary of Commerce for Science and Technology plans a new civilian R&D initiative. His office is working on a proposal to establish a Cooperative Technology Program, under which the Federal government would help finance the development of basic technologies of value to an entire industry or several industries. Similar proposals in the past have been geared to rescuing ailing industries, but one of the options under consideration contemplates joint government-industry efforts to identify and exploit technological opportunities in leading sectors of the economy. (6)

The point of mentioning these anomalies is not to criticize
current policies but to suggest that the nature and extent of the
government's role in civilian R&D are hardly closer to being
settled than they were in previous administrations. (7) While
impressive in the aggregate, the government's involvement represents a series of piecemeal responses to specific social needs
and perceived opportunities. In some cases, Congress created an
R&D program for lack of any better alternative; the appearance
of trying to solve a problem can assume as much importance as its
accomplishment.

This pattern may be changing, however, with the emergence
of three general concerns. One concern mentioned earlier is
that civilian research, development and demonstration programs,
in contrast to military and space activities, have not been
highly successful in producing important innovations. (It should
be noted that there is doubt among economists even about the
spill-over benefits of defense and space R&D expenditures. (8))
As Frank Press commented in a 1977 Science editorial, "Its impact
on meeting public expectations -- on filling the everyday needs
of people -- often seems disappointing." (9) And a recent
report for the OTA observed, "Federal expenditures for demonstration projects . . . have grown to over $1 billion annually, and
further growth appears likely. Yet their effectiveness has been
limited." (10) This growing insistence that R&D prove beneficial
is principally a result of energy concerns and constraints on the
Federal budget as a whole. It may be that we lack a systematic
evaluation of R&D programs or that the results simply reflect
the riskiness of R&D in general; but these qualifications are not
very persuasive when it comes to the expenditure of public funds
on urgent national problems, especially when the experts are
generally critical of the government's performance.

Secondly, there is concern that the innovation which both
private and public R&D are supposed to fuel is seriously lagging
and that the failure of American firms to market more new products
and institute new manufacturing processes is responsible in large
measure for the nation's sluggish economic and productivity growth
and declining trade competitiveness. Among the various indicators
that have been cited as evidence of this trend, two were singled
out as most disturbing by witnesses before the Senate Subcommittee
on Science, Technology, and Space:

> ●There has been an astonishing drop in the
> creation of small high technology companies
> which in the past have been responsible for
> introducing a disproportionate share of
> innovations. Several years ago, two or
> three hundred venture companies entered the
> market with new issue underwritings each
> year; in 1977 there were 46.

●Existing firms in R&D-intensive industries
have transferred some investments from
major new product and manufacturing innovations to relatively minor product and
process improvements promising short-term
returns. Although total industrial R&D
spending has somewhat more than kept pace
with inflation, there apparently has been
a significant shift from research to
development as well as a decline in investments in new plant and equipment which may
incorporate new technology. Admittedly,
the shift has been hard to quantify, in
part because the conventional R&D categories do not apply as readily to industry
as government; but the impression is
widely shared that it is occuring and even
accelerating. (11, 12)

Finally, many observers are alarmed that our chief foreign
competitors are investing increasing shares of their GNP's in
research and development and a far higher proportion of government R&D in the civilian sector and that they are reaping hand-
some returns in productivity gains and exports from these investments. Total U.S. expenditures on R&D have declined from 3% of
GNP in 1964 to about 2.3% in 1976, in contrast to increases in
Japan and Germany in the same period. According to OECD figures,
36% of U.S. government R&D funds in 1975 were spent on economic
development, energy, health, community services, and the advancement of knowledge, compared with 92% in Japan, 85% in Germany,
and 65% in France. (13) These governments are playing a direct
role in the development of major technologies such as computers
and electronic devices as well as aviation and nuclear energy by
supporting networks of industrial research institutes, cost-
sharing arrangements and other means, although they have also
vigorously pursued a "market pull" strategy through government
procurement, tax incentives, loans, manipulation of market structures, and provision of capital to new venture companies. (14)
We cannot assume that there is a causal relationship between our
competitors' publicly supported research and development programs
and their superior trade performance, but neither can we afford
to assume that there is none.

In short, concerns in Congress, the Administration, and the
private sector about the productivity of Federal R&D go beyond
the achievement of specific goals to the state of American industrial technology in general. For the first time, therefore, there
is a basis and some urgency to address the role of government R&D
as a whole, as well as the effectiveness of R&D in transportation,
energy, agriculture, health and so on. The issues are both
complicated and controversial: To what extent is the economic

climate so adverse to innovation that R&D results are stymied
regardless of their source? Can the climate be improved by
removing disincentives, creating new incentives, or by changes
in the structure and behavior of the private sector? To what
extent can public investment in R&D stimulate innovation? In
what circumstances does it drive out private investment or, for
that matter, contribute to overinvestment? What should be our
goals and where can the government be effective?

A concerted effort to answer some of these questions entails
certain risks. The debate could drift aimlessly, as similar discussions have before, leaving industry, government, labor and
public interest participants more skeptical of one another's
motives. Furthermore, the lack of public awareness underscores
the magnitude of the task of leadership. The importance of
innovation to the national welfare has yet to capture the attention even of the Administration's economists, let alone the
imagination of the public. On the other hand, there are officials in the Executive branch and members of Congress willing to
assume leadership. The private sector is lending its support
to both the domestic policy review and congressional inquiries.
The circumstances will not improve if the present opportunity is
missed.

Managing Government and Industry Collaboration

What may be less apparent is the simultaneous opportunity
to resolve the so-called "operational" problems of civilian R&D
management, many of which are not so far removed from issues of
high level policy as some might suppose or often wish.

Within the past few years, a substantial research effort
has been mounted to identify the factors associated with success
or failure in implementing the results of Federal research. This
literature includes the House and Jones study (15), the A. D.
Little report for ETIP in 1976 (16), the Rand Corporation analysis
of Federally funded demonstration projects in 1976 (17), and a
recent study for ETIP, Management of Federal R&D for Non-Federal
Applications, by the Stanford Research Institute (18). The SRI
report is based on a quantitative analysis of data obtained from
interviews with agency officials, R&D performers, and potential
beneficiaries of 46 projects in various programs of eleven
Federal agencies. Its appearance is an appropriate occasion to
take stock of the accumulated findings.

At the risk of over-simplifying, SRI's results, which are
presented as a set of guidelines for project management, generally
confirm the thrust of pervious studies. Projects should be selected on the basis of user needs and designed to accommodate
market uncertainties. Commercialization is much more likely to
occur if stated and agreed to as a goal. It is important for the
agency and R&D performer to cooperate in developing a deployment
strategy from the beginning of the project. Communication among

managers, performers, manufacturers and users is essential. Cost-
sharing can increase the stakes in cooperation and thus improve
the chances of success.

So much of this borders on common sense that one wonders
why these procedures are not more routinely followed. Yet time
and again, SRI found that the factors regarded as crucial by
government respondents were not predictive of success and that,
generally speaking, ". . . Civilian agencies with R&D programs
destined for the non-Federal sector have not been following R&D
management practices that, if followed, would lead to greater
commercialization results." (18)

Several propositions are implicit in the SRI and other
studies but need to be emphasized. In order to be of benefit,
civilian research and development must serve the purposes of
industry as well as government. Its conduct and outcome are
shaped by the values, interests and professional perspectives of
all of the participants. Some of these interests may be cross-
cutting; for example, program directors and industry executives
are likely to be concerned with end results, researchers and
engineers with technical sophistication and success.

By far the most problematic relationship, however, is that
between government and industry. The government seeks to spread
economic benefits, the private firm to capture them. The
Federal agency may desire a major technological advance while
the manufacturer is inclined to risk marketing only an incremental one. Industry generally resists external interference in
the later-stage development and marketing decisions which are
thought to be crucial to a project's success but all too often
ignored by R&D decision makers. Ironically, such incongruities
between public and private interests, which cloud the prospects
of commercialization, are frequently the very reasons for government intervention.

The problem, therefore, is not simply that profitability
does not govern public decisions. Administrators must satisfy
a large number of constituencies, including but not limited to
producers and consumers of particular goods and services. Public
authority is highly fragmented among committees of the Congress,
agencies of the executive, and various levels of the bureaucracy.
Complex procedural constraints reflect the traditional tensions
between government and the private sector. For example, program
and project directors cannot be expected to consider regulatory
or other incentives to commercialization of R&D if such instruments are outside their responsibility or their agency's authority. They are usually bound by procurement procedures developed
for the government's own missions. They are frequently restricted
in granting property rights and in setting up advisory committees.
And increasingly, they are subject to organizational conflict
of interest rules which may discourage advocacy by industry and
continuity of collaboration with particular firms in the name of
objectivity and competition. (19)

Government needs to be better educated in the realities of the marketplace; but even in civilian research and development, its actions cannot be guided solely by them. Nor is the reconciliation of government and industry interests simply a matter of consulting one another. If the previous hypotheses are correct, what is required is the institutionalization of private sector participation in public policy decisions and management. This proposition is radically at odds with the more extreme versions of the "hands off" philosophy of some executives in industry and the "arm's length" philosophy of some officials in government.

"Institutionalization" does not mean the establishment of permanent relationships between agencies and firms or industries. R&D for commercialization implies a limited government intervention and eventual withdrawal. Nor is it necessary, even if it were possible, to establish elaborate R&D and dissemination networks such as the agricultural extension service. Rather, the task is to formalize procedures and ground rules for negotiating limited collaboration among government, industry and universities for specific mutual goals, facilitating reconciliation of interests that are at odds, and protecting the public interest in preserving competition.

While this is no simple task, the development of criteria for Federal civilian R&D investment, and by implication, non-intervention, is a longer term effort; indeed, the latter is an evolutionary goal. Yet the institutionalization of private sector participation in R&D programs would facilitate the flow of information and counsel from industry that is needed to inform decisions about where, under what circumstances, and to what extent the government ought to commit its resources.

A recent Office of Technology Assessment report points out the opportunity to develop such procedures under the 1977 Grant and Cooperative Agreement Act. (20) Whatever the vehicle, it will not happen automatically. Nor will the sensitive political issues of R&D management be addressed adequately in the context of particular programs when they are affected by governmentwide norms and policies. Unless civilian R&D efforts are perceived to have a bearing on the nation's economic problems, it is likely that policies which are inimical to the requirements of commercialization will be adopted, debates will be prolonged over such matters as government patent policy, and fortuitous opportunities such as the advent of cooperative agreements will be missed. In short, the operational problems of R&D management should be a prominent part of the national discussion of industrial innovation policy.

LITERATURE CITED

1. Schnee, Jerome E., "Government Programs and the Growth
    of High-Technology Industries," Research Policy (1978),
    7, pp. 2-24.

2. U.S. Office of Management and Budget, "Special Analyses:
    Budget of the United States Government," 306-07,
    Washington, January 1978.

3. U.S. Senate Subcommittee on Science, Technology, and
    Space, "Symposium on the Future of Space Science and
    Space Applications," 40-55, Washington, 1978.

4. The President's State of the Union Message, January 1978.

5. Peterson, Russell W., "The Role of R&D in Meeting
    Societal Objectives," Remarks to the R&D Colloquium of
    the American Association for the Advancement of Science,
    Washington, June 20, 1978.

6. U.S. National Bureau of Standards, "What Will the Nature
    of the Cooperative Technology Program Be? Issues and
    Answers," Prepared for the Office of the Assistant Sec-
    retary for Science and Technology, Department of Com-
    merce, Washington, April 17, 1978.

7. Eads, George, U.S. Government Support for Civilian
    Technology: Economic Theory Versus Political Practice,"
    Research Policy (1974), 3, pp. 2-16.

8. U.S. National Science Foundation, "Preliminary Papers
    for a Colloquium on the Relationships Between R&D and
    Economic Growth/Productivity," Washington, November 9,
    1977.

9. Press, Frank and George Busbee, "Intergovernmental
    Science and Technology," Science (May 27, 1977), 196.

10. U.S. Congress Office of Technology Assessment, "The Role
    of Demonstrations in Federal R&D Policy," ix, Washington,
    July 1978.

11.   Mogee, Mary Ellen, "Industrial Innovation and Its Rela-
      tion to the U.S. Domestic Economy and International
      Trade Competitiveness: Analysis of Hearings Held by
      Subcommittees of the Senate Committees on Commerce,
      Science, and Transportation; and Banking, Housing and
      Urban Affairs; and the House Committee on Science and
      Technology," Congressional Research Service: Washington,
      October 13, 1978.

12.   Shapley, Willis and Don I. Phillips, "Research and
      Development: AAAS Report III," 53-71, American Asso-
      ciation for the Advancement of Science: Washington,
      1978.

13.   U.S. National Science Board, "Science Indicators 1976,"
      5, 186-7, Washington, 1977.

14.   M.I.T. Center For Policy Alternatives, "Government
      Involvement in the Innovation Process," 59-68,
      Office of Technology Assessment: Washington, 1978.

15.   House, Peter W. and David W. Jones, "Getting It Off
      the Shelf: A Methodology for Implementing Federal
      Research," Westview Press: Boulder, Colorado, 1977.

16.   Little, Arthur D., Inc. "Federal Funding of Civilian
      Research and Development," Prepared for the Experi-
      mental Technology Incentives Program of the National
      Bureau of Standards, Washington, 1976.

17.   Baer, Walter S., et al, "Analysis of Federally Funded
      Demonstration Projects," The Rand Corporation: Santa
      Monica, California, 1976.

18.   McEachron, Norman B., et al, "Management of Federal R&D
      for Commercialization," SRI International: Menlo Park,
      California, 1978.

19.   Rawicz, Leonard, "Organizational and Individual Con-
      flicts of Interest: Impact of the Rules on Contract-
      ing," Paper presented to the Federal Bar Association/
      Bureau of National Affairs Conference on Government
      Contracts, Philadelphia, March 13-14, 1978.

20.   U.S. Congressional Office of Technology Assessment,
      "Applications of R&D in the Civil Sector: The
      Opportunity Provided by the Federal Grant and Coopera-
      tive Agreement Act of 1977," Washington, June 1978.

RECEIVED March 14, 1979.

Can You Innovate in Uncle Sam's Embrace?

ARLEN J. LARGE

Wall Street Journal, 1025 Connecticut Ave., NW, Washington, DC 20036

There currently is little dissent from the discovery of an alleged new National Problem: a decline in industrial innovation in the United States, stemming from the asserted reluctance of American companies to perform basic research on their own, or even use much of the existing research data already financed by the government. The Carter administration has been considering what Assistant Commerce Secretary Jordan Baruch calls "a wide range of tools with which to motivate the private sector's behavior with respect to the rate and direction of the innovation process." (1)

The official momentum in this area is quite high. As the bureaucratic and legislative machinery cranks away in the years ahead, something actually may come of it; indeed, some limited good may come of it.

But industry's research and development managers should be fully aware of the potential costs that might go with being "motivated" from Washington. The wide range of tools could include some monkey wrenches in their labs. There ought to be second and third thoughts before America's private businesses become more closely entwined with government in the pursuit of claimed national goals, as is the case in, say, Japan. There is still a lot to be said for maintaining a cool, correct, arms-length relationship between the worlds of business and government.

As this subject is considered, some basic points should be kept in mind:

0-8412-0507-8/79/47-105-015$05.00/0

--The burden of proof must always be borne by people who claim to see a "national problem" requiring government action. Government action often means government favors to those who can put an attractive label on their problem. "National security" is an old favorite, used, for example, by the oil companies in the 1950s and 1960s to justify import quotas to protect their domestic prices. "Innovation lag" is the latest label in search of a favor.

--People who get the favors must pay a price. Tax incentives or loan guarantees intended to stimulate more private R&D are a guaranteed source of new government regulation that businessmen already blame for stifling the innovative impulse. Changes in patent policy, desirable as they may be, could open up a whole new area of government rule-making that mainly activates the technology of lawyers' cash registers.

--The innovation partnership, the bestowing and receiving of favors, will increase the already-unhealthy amount of attention that business pays to government. Uncle Sam can never be ignored, of course, as long as he chooses to bankroll so much of the nation's R&D effort, but the scientific knowledge underlying all industrial innovation shouldn't be allowed to become totally dependent on the political process in Washington.

The Unchallenged Concensus

Science and technology policymakers in the Carter administration have taken the lead in decrying a decline in U.S. industrial innovation and a reluctance to commercialize government-financed research. Business executives, sensing something beneficial in the works, have eagerly joined the lamentation. The alarm often seems to exist at the bumper-sticker level of analysis, reflecting a chauvinistic fear that Americans will have to stop shouting "We're Number One" in world technology. Some people can't seem to bear the thought that many American consumers stubbornly prefer to buy their color TV sets from Sony than from a U.S. maker. There are warnings, voiced in semi-protectionist terms, about a loss of markets and jobs to foreign geniuses unless something can be done to re-inspire good old Yankee know-how. There now is a fairly solid consensus that the Federal government should "do something." That consensus has been little challenged so far.

Yet it should be. Proponents of government action to
stimulate private innovation have the burden of proving that
the cost of the subsidies, invisible though they may be, will
return value to society as a whole. That burden, I believe,
has not yet been met. In its first annual report to Congress
on science and technology in 1978, the National Science
Foundation ducked questions about what the government should
do to stimulate innovation. But its assessment of the U.S.
technological position relative to other nations didn't sound
much like the after-dinner speeches of government and industry
officials who have been scaring everybody about the U.S.
innovation lag. Said the report:

> "Neither the available economic nor technical indicators
> provide hard evidence of an eroding U.S. technological
> position which can be tied to negative economic conse-
> quences. The U.S. continues to gain strength in the com-
> mercial exploitation of technology. However, in specific
> technical fields some foreign competitors likely will
> overtake the U.S. and some will fall further behind.
> Overall, however, the U.S. appears to be maintaining
> its scientific and technological advantage." (2)

The NSF report also discussed one supposedly ominous
symptom of the innovation decline, the well-publicized shift
of private funds away from "R" and into "D." The NSF found
that such a shift has indeed been taking place, but it added
this inconclusive note: "At present there is no general agreement as to whether this trend will have an adverse effect on
the U.S. economy." (3)

A Board of Directors Problem

Well, if it doesn't hurt the U.S. economy, it is not a
national problem. It is a board of directors problem at each
individual company in the nation. The after-dinner speaker's
standard lament that "we" are letting the world's technological
lead slip away glosses over the continued outstanding performance of many American companies. Policymakers at these
firms have their eyes on sources of profit five and ten years
from now, and are salting away money in the requisite lab
facilities and research PhDs, instead of announcing nice
immediate dividend increases that would please Wall Street.
Society generally need have little sympathy for other companies that aren't equally farsighted about what it will take
to survive in the marketplace in the 1980s. Let them go under.

It is no business of the government to guarantee the success
of uncompetitive business.

Businessmen often say in their defense that it's increasingly difficult for boards of directors to make long-
range R&D commitments because of uncertainty about future
government regulations. That is doubtless true, but efforts
should be made to correct it regardless of any feared innovation decline. Government regulations need to be as pre-
dictable as possible, and applied with more common sense
than heretofore, whether or not there's an energy crisis or
a scarcity of venture capital or an innovation lag. These
popular focal points of complaint shouldn't be the main
justification for making government regulators meet their
prime responsibility: doing a better job in the first place.

Tax Rewards

The main objective of taxation should be to raise enough
revenue to meet the government's needful expenses. But both
Republican and Democratic administrations in recent years
have shown an unfortunate tendency to use the revenue system
as a means of rewarding government-approved behavior. You
get a tax reward by owning a home, receiving stock dividends,
contributing to churches and politicians, installing a productive new machine, and fighting the energy crisis by buying
a storm door, instead of a frowned-upon pool table for the
basement.

Thus by now it's almost an automatic reflex for government policymakers to think about stimulating innovation
by dangling the reward of tax credits or fast depreciation
writeoffs. The rewards, of course, would be narrowly "targeted"
to cover added R&D investments. Targeting is a key concept
of government officials who are trying to manipulate behavior,
because without it, a tax reward would become a "windfall"
for everyone, innovative or not.

If an innovation tax reward was on the books, probably
a majority of companies would see their interests coincide
with those of the government, and would dutifully build new
labs and buy new research equipment. Perhaps some others would
be tempted to veer a bit off target and sink the innovation
reward, or its bookkeeping equivalent, into a new employee
cafeteria or a fleet of trucks. And, unfortunately, a few
businessmen doubtless would try to figure out ways to just

take the money down to Florida and spend it on sin.

There would be just enough of that to warrant the establishment of a new Internal Revenue Service team of specialists to
ferret out abuses of the innovation tax reward. Just as there
are reams of IRS regulations spelling out what kinds of new manufacturing equipment qualify for the existing investment tax
credit, there would be new bundles of rules attempting to define
what is qualified "development," and to identify the elusive
point where it shades off into unqualified "production." What
is now done casually and easily around the lab would begin to
conform to the rigidities of the Internal Revenue Code. R&D
managers, though grateful for the extra money, might begin to
wonder whether it's really worth the new hassle.

The Slippery Slope

Federal loan guarantees would be another kind of tool for
stimulating innovation. Congress has already started down that
slippery slope by authorizing loan guarantees for demonstration
plants using new Federally researched techniques for coal gasification, for example. This is a key step along the way to
commercialization of the vast effort that the government is
pouring into energy R&D, but the increasing use of government
loan guarantees would drastically change the way industrial
innovation has been financed in the United States. The mere
prospect that a loan guarantee might be available would tend
to make banks turn a cold face to a risky project until a
guarantee actually comes through. And whether it does or
not could depend less on the technological worth of the project than on the influence of the U.S. Senator from the state
where it's to be undertaken. A company benefitting from a
guaranteed loan would find itself in thrall not just to the
bank making the loan, as always before, but to a new set of
masters in the Treasury whose duty is to protect the government's
interests.

Congress has explicit Constitutional authority "to promote
the progress of science and useful arts" by establishing patent
monopolies for inventors. Changes in patent policy offer perhaps the government's most promising and ungimmicky way of overcoming whatever innovation lag may exist. But change should
be kept simple, such as a straightforward increase in the life
of a patent from 17 years to 25 years. Regrettably, if the way

Congress writes tax laws is an example, complexity will creep
in. There may be great merit to proposals, for example, to
give the Patent Office, instead of the courts, a greater degree
of finality in issuing a patent, but that would make a bill
harder to understand and harder to pass. It should be kept
in mind that Congress hasn't changed basic patent law in
many years, and it has no institutional memory of how to
do it.

Labs, Not Lawyers

A special problem is how to deal with the private patent-
ability of discoveries arising from government-financed research.
The current mish-mash of rules enforced by different Federal
contracting agencies is thought to be a serious barrier to
commercialization of products and processes. It's not the purpose here to propose solutions to this problem, but merely to
urge again that the "reforms" be kept simple enough to understand without the aid of a lot of lawyers. A company which
gets an exclusive license from a university that has patented
a government-financed discovery should have confidence that
it won't have to transfer budget resources from its R&D lab
to its legal department just because it got involved, however
indirectly, with Uncle Sam.

Unfortunately, it's getting easier all the time to become
involved with Uncle Sam. The government presses its attentions
in a growing number of areas as Presidents and Congressmen seek
election as "problem-solvers" and then seek problems to solve.
Given the nations's political and economic traditions, the
only practical way Washington can induce many problem-solving
activities is through contracts with private institutions.

In the name of national security, contracts become aerospace
industry payrolls. In the name of the energy crisis, contracts
try to create whole new fuels industries that don't now exist.
In the name of supporting basic science. contracts put bread
on the table for hosts of university researchers. For all these
people and many more, government-paid problem solving has become a livelihood, for the most part eagerly sought. Thus has
the Federal contract dollar permeated the entire U.S. economy.

But there is a contradiction here. Most people doubtless
would oppose the idea of extracting money from the pockets of
the taxpayers for the enrichment of private industry, if you

put it that way.   That, of course, is the reason for the hang-
up about transforming government research money into monopoly
patents that might create private profit.   Embarrassment about
that produces all those tangled rules that the Federal agencies
and universities lay down.   The government likewise resists the
idea of just mailing a check to a private corporation for construction of its new R&D lab, and the company's president would
faint dead away at the thought of being photographed standing
in line to cash it at the bank.

Beware the Enforcers

Thus the problem of the innovation lag must be solved with
less visible tax dodges, or the vague promise of a tradeoff between saving money on air pollution controls and spending it
instead on new product development.   But this inevitably will
require the deeper and deeper involvement of IRS enforcers and
regulators to see that the innovation lag problem is indeed
being solved through these indirect stimulants without everybody running off with the money.

At the heart of the contradiction between contract/tax-
dodge problem-solving and the taboo against private enrichment
from the Treasury is the concept of targeting, the promise of
a benefit only if you behave in an officially prescribed way.
To the extent that their political impulses will allow, government office-holders should lay off trying to target their benefits.   Instead of a tax credit only for companies that raise
their R&D investments, Congress should just put through another
general cut in corporate tax rates.   That way there would be
no need to send the IRS man around to count the test tubes in
the new lab.   A general tax cut recipient would be free to use
the extra money to install a new executive dining room, squander
it on dividends or use it to develop profitable products for
the future, with the marketplace ultimately deciding who did
the right thing.

Inventors and would-be entrepreneurs have especially com-
plained about capital gains taxation as a drag on their ability
to attract venture capital.   While it opposed a general cut in
capital gains taxes, the administration included officials
who were sympathetic to a capital gains formula skewed to benefit small companies that might strike it rich with a hot new invention.   That's another example of targeting.   One may quarrel
with the final decision of Congress in 1978 to go ahead with a
general capital gains tax cut, thereby deepening the class

distinction between different kinds of income, but surely that
is preferable to a complex and discriminatory plan that would
reward some investors but not others in conformity with some
supposed national policy goal.

It's true that if govenment policymakers stop trying to
target benefits to solve problems, they would be less able to
take credit for finding problems and proposing 17-point "solutions." But if the office-holders who are trying to overcome innovation lag would just do something else for the next
ten years, they might come back to find that the problem has
solved itself.

The Real World

America's great attractiveness is its diversity. People
still are able to go about their business and pursue their
own interests without the cadence of a single drummer. The
nation-state still weighs less heavily here than elsewhere.
It's a common business complaint that this is changing, that
the government is intruding too much, that the IRS is too
nosy, that HEW demands too many forms, and that the OSHA in-
spector is about to break down the door.

That may be true, but the intrusion is mutual. Washington
is filling up with government affairs departments of corporations
whose agents persuade their bosses that their presence in the
capital is a requirement of the "real world." They have been
following with keen interest--too keen--the administration's
innovation study, hoping to be able to report possible tax
breaks and softer regulation to the home office. The proffered
"partnership" for solving this asserted new national problem
thus promises to smudge further the dividing line between the
interests of business and government, with the seemingly eager
consent of both parties.

Perhaps it would be more cost-effective if business were
to ignore the government a little more than it does now, bring
its spies home from Washington and put more of its own chips
on innovative technology for future profit without waiting for
an official reward. Washington is not the "real world." The
real world is nature, and it's always out there waiting to be
manipulated for man's benefit, with or without guidance from
the government.

Literature Cited:

1. Interview, Research Management, September, 1978
2. National Science Foundation, "Science and Technology
   Report, 1978," Committee Print, Committee on Science and
   Technology, U.S. House of Representatives, Government
   Printing Office, page 84.
3. Ibid, page xiii.

RECEIVED March 14, 1979.

Federal Policy Concerns Regarding Commercialization of Federally Funded R&D

RICHARD PENN

ETIP, Center for Field Methods, National Bureau of Standards, Department of Commerce, Admin. Bldg. Rm. A-740, Washington, DC 20234

I am pleased to have the opportunity to discuss the programs that the office of the Assistant Secretary for Science and Technology is currently pursuing with respect to the concerns in the Federal Government about the commercialization of Federally funded R&D in industrial innovation. I should like to discuss first of all some of the general dimensions of the problem as is seen from the Department of Commerce and then to discuss briefly a number of programs that are underway or planned within the Department. These include the activities of the Center for Field Methods or the Experimental Technology Incentives Program (ETIP) at the National Bureau of Standards; activity with respect to the Domestic Policy Review on Industrial Innovation; the program of the Department with respect to aiding impacted industries; plans with respect to a cooperative technology program; and finally, activities which the Department is pursuing that address commercialization of Federally funded R&D, patent policy with respect to inventions and patentable activities flowing from government funded research and development.

Why should the Department of Commerce be concerned with commercialization of Federally funded R&D? I think the statistics with which we are all familiar amply point to problems with a declining rate of innovation and use of technology in this country. One need only examine trade statistics, the rates of patents being sought by U.S. inventors, the levels for support of research and development to have some idea of why the government is concerned with the rate of technological change. Economists have now established with considerable certainty that technological change is one of the most important contributors to the growth of productivity in this country. We

perceive that the unit of change when one is talking commercialization or industrial innovation is the firm. It is at
the level of the firm that decisions are made to incorporate
and use new technology in its products and processes. We
are also convinced that due to the variability of the various
industrial sectors, in many important ways, it is important
that government policies which are designed to be effective
be sector specific. We are well aware of the public policy
problems that this creates when viewed against the traditional
equal treatment under the law philosophy which has pervaded
in this country since its inception. We also can see that
the innovation problem cannot be cured by technology policy
alone. Decisions to innovate are influenced by a whole host
of economic and business variables which must be treated if
government policy is to be effective. Technical answers are
a necessary but by no means sufficient part of promoting
greater use of technology by the private sector.

We do perceive that there are a number of rationales for
government involvement in the innovation process. These include the traditional economic argument of imperfections in
the function of the marketing. Further, there is market frag-
mentation. We have a situation where information does not
perfectly flow from one portion of the market to another.
And thirdly, we now have competition on the international scene
from combinations of firms and nations with which it is very
difficult for industry in this country to compete.

We need not argue whether or not government should enter
into the play of the market place. It is already there. One
of the things that we need to do is to try and find ways to
remove the traditional adversary relationship which has governed
many interactions between government and business and become to
realize that it is not my problem or yours but it is ours. We
need to collectively work towards solutions of our mutual problems for the economic health of the country. We need to find
ways for making institutional change so as to get an appropriate cooperative atmosphere such as that which prevailed in
the country during World Warr II when industry and government
harnassed its collective know-how to develop substitutes for
natural rubber which was denied to us during that conflict.

I must point out that we conceive the process of innovation
as really being a non-process but rather being a series of rather
disjointed events which began with ideation, then followed by

invention, development, demonstration, marketing and diffusion. We conceive that there are currently three types of technology designed to serve different purposes which the government must interact with in different ways. First, there is research and development which is funded for the end use of the government itself. That is, primarily, space and defense research leading to hardware procured by the government. Here the problem is how to appropriately provide for spin-off or transfer of the technological lessons learned in the commercial marketplace.

Secondly, there is government funded research for social problems such as air pollution. In this case, the traditional economic argument for government involvement is the non-appropriability of sufficient benefits of the research to any particular segment of the business community to persuade it and/or provide an adequate incentive for it to fund the research.

And finally, there is research and development for purely commercial purposes. Where the government may elect to participate in the R&D process in order to assist American firms in attaining and retaining an appropriate competitive position vis-a-vis firms in other industrialized countries.

Let me further observe that it is no news to this group but maybe to others that the so-called science and technology policy of the government at the moment is a non-policy, that is, there is no over-all coordination of a general policy of the government towards science and technology. It is rather administered in a fragmented manner by the several departments and agencies in the executive branch further influenced by the congress and rulings of the courts.

So much for the general setting of the activity of the Department of Commerce. Now let me turn to the specific programs which are operated under the general cognizance of the Office of the Assistant Secretary for Science and Technology.

The Experimental Technology Incentives Program is now a portion of the Center for Field Methods at the National Bureau of Standards. It was instituted in 1972, when the science advisor articulated many of the same problems that we see being raised in the press today. The program has sought to understand the interaction between government policy and practice

and industry ability to innovate in four specific program areas.
These program areas include government procurement policy, government regulatory activity, government economic assistance, and
government direct funding of research and development.

The procurement intertests of the ETIP program are currently looking at how one articulates the government needs in
terms, usually conceived as being performance requirements,
such that a maximization can be achieved with respect to the
ability of industry to innovate. The government can and has
acted as an initial market for innovative products when it
has been willing to accept the risks associated with buying
new and only partially tested items for its own use. We are
further interested in how the government in its procurement
activities for research can organize and utilize the full
talents of the marketplace in defining the best products to
meet its needs. Particularly this is important in carrying
out government funded research where no single institution
possesses the requisite knowledge of research and the marketplace that is necessary in order to optimize that procedure.
Our experience here has been typified by the funding of a research consortia to carry out flammability testing with respect
to cotton, polyester, blends in apparel and clothing where the
consortia directed the activity in a more coherent and desirable fashion than would a research program directed by
single segment of the community such as the university or the
business firms.

In the area of regulation, and we hear no other topic discussed so frequently by business as being under a constraint
in their ability to innovate. Our efforts have been directed
towards eliminating some of the uncertainty and attempting to
cut the delays which are inherent in the current practices of
the regulatory process.

With respect to economic assistance, we are exploring with
the Securities and Exchange Commission the rules under which
they provide venture capital to small firms. Recent hearings
have shown that these rules are unduly restrictive and expensive
with respect to the ability of small high technology firms to
raise equity capital. As most of you know, equity capital is
a fundamental method of funding small high technology firms. And
if the Securities and Exchange Commission can find ways to improve the ability of these firms to raise money while still
providing adequate protection to the private investor then considerable progress will have been made in this important matter.

With respect to the direct funding of research and development, Steve Merrill earlier discussed the research findings of Arthur D. Little, the Rand Corporation, and SRI International, with respect to trying to learn from past government activities which items have been successful and which have not resulted in commercialization from government research directed at commercialization. We will actively seek in the next year opportunities to persuade government agencies to try in a prospective mode those guidelines which have been provided by these three studies.

Now let me turn to the Domestic Policy Review on Industrial Innovation. This Review was directed by Stuart Eisenstat on behalf of the President on May 9, 1978. Contrary to statements made earlier in this session, we do not view the Domestic Policy Review as having reached a conclusion on April 1, 1979. Rather, on that date we will have presented to the President certain options which he can exercise with a view of improving the situation with respect to the ability of firms to innovate, but there will be required careful attention to the implementation and evaluation of those options as a follow-on to the Presidential decision. Further, we expect that there are likely to be a number of areas identified where consensus cannot be reached and further research will be necessary before decisions can be made with respect to the undertaking of certain steps. At the present time, the advisory committee structure in industry is taking shape which will act as an input to the entire process by the industrial sector. The Economic and Trade Advisory Panel under the aegis of Bill Agee of Bendix has already been formed. The Panel on Regulation and Environmental, Health and Safety, under Don Frey of Bell and Howell, will meet shortly. Other panels are being convened on Regulation of Industry Structure and Competition, on Patent and Information Policy, and on Federal Procurement and Direct Support of R&D. The meetings of these panels will all be public. These will be followed by public seminars at which their recommendations are discussed. We expect from this over-all exercise that there will emerge a number of carefully tailored, sharply focused options which the President will direct to come into being. We foresee at this time that the Presidential actions can fall into three categories. Those which will require legislation in which it will be necessary to enlist the aid of the Congress. Those which will require Presidential directive in the form of executive orders and

those where the President can request or direct agency heads to
make use of existing authority already in place to carry out the
necessary activities.

Now let me turn to the discussion of assistance to impacted
industries. Here the Department of Commerce Science and Technology organization is assisting the Economic Development
Administration in providing the technical component of assistance to industries which have been impacted by foreign competition. An example is the U.S. shoe industry, which has currently fallen well behind our foreign competitors in the point of
sales. Here, in consultation with industry, there have been
identified certain technological needs which the industry, because of its fragmented situation, and in the absence of research conducted on its own behalf, has been unable to meet. On
a cooperative basis between government and industry, we are in
the process of attempting to find technical solutions which will
not interfere with the normal functioning of the marketplace, and
which can be used to help the industry in such areas as the
molding of women's shoe bottoms. There the style changes mandate rapid changing of molds, but the supply process for in-
jection molds now requires a lengthy and extended time to develop new molds.

Let me now turn to our concept of a Cooperative Technology
Program. Here we anticipate a program whose details will be
worked out in the course of a year long study which has been
funded for fiscal year 1979. This program will look at the creation of infra technology in a way that marries the realities of
federal policy development with our knowledge of the innovation
process and the needs and peculiarities of specific industries.
We hope that this program will be a significant contributor to
the promotion of U.S. industrial competitiveness, productivity
and profitability and, thus, will be an aid to the social and
economic well-being of the nation.

As conceived at the present time, the program could focus
on trade impacted industries, such is currently being done in
our impacted industries work. But also importantly, it could
stretch to lead industries, particularly those threatened by the
high R&D investments in foreign countries; and to industries,
which when stimulated by technological development, will have a
potential for social benefits such as energy conservation, new
employment, and new apparatus for regional development.

Based on proposals from industry, we would enter into infra-technology development programs jointly staffed by appropriate technical people from government, industry and the academic community.

As currently conceived, each project would have the following steps in it: 1) identification by industry of problems and opportunities; 2) identification and analysis in collaboration with industry of technological solutions and of probable impacts, technological, economic, industry, and social; 3) execution of research, development, and technological tasks by government, cooperating industries, and selected industry R&D support capabilities; 4) monitoring and evaluation of the project process and activity; and 5) the orderly termination and phase-out of federal involvement and the transfer of the development and marketing to the commercial marketplace. The concept is clearly in mind, but of course, many details need yet to be worked out.

And finally, let me turn to the problem which has already received a great deal of attention at this session. That of the status of patents which are developed out of government sponsored research and development. It is worth recalling that in its final report in December 1972, the Commission on Government Procurement concluded that government patent policy could have the most significant effect on technological innovation. To-day, the problem of ownership of government patents is being addressed by a high administration panel, chaired by Dr. Jordan Baruch, the Assistant Secretary for Science and Technology in the Department of Commerce. The body is the Committee on Intellectual Property and Information. The committee is struggling with the very difficult problems. Initial response of all the federal members has been very favorable towards the propositions set forth and this includes both the Department of Defense and the Department of Justice.

There is general agreement that there exists a serious problem, and it is important for the full committee members to participate personally in working towards a solution. The committee operates on the principle that technological innovation, that is, the development of new inventions, is a primary means for achieving non-inflationary economic growth, job creation, and a stronger international position for America and American industry. Technological innovation also promotes competition within the economy. Government policy with respect to the allocation of rights and patentable inventions resulting

from federally supported research and development by non-
governmental persons bears a major responsibility to the pace
of technological innovation in America today.

It is also fair to say that there is wide support for a
proposition about how a desirable government patent policy
should be carried out. We hold that government patent policy
should strive to: (1) obtain the best contractor effort for the
government; (2) maximize technological innovation; (3) promote
competition within the private sector; (4) recognize the public's
equity in the products of federally supported research and
development; and (5) strengthen the research programs at
universities. In addition, we believe the government patent
policy should: (6) be uniform in the sense that similar cases
should be treated similarly no matter which government agency
provides the support and there should only be a single set of
patent regulations with which a potential government contractor
must deal; 7) we must be flexible in the sense that different
cases should be treated appropriately, that is not necessarily
identically; and (8) the systems should be as clear and simple
as possible.

We are certain that everyone will agree with these eight
general characteristics of a desirable government patent policy.
But I am certain that everyone recognizes that achieving all of
those goals in a balanced way will be most difficult.

I have tried to outline our concepts of the problem in
industrial innovation in the United States to show some of
our concerns about the problem and to set out a rationale of
why the government should be involved. I have then attempted
to describe very briefly to you an overview of the five programs that we have underway under the aegis of the Assistant
Secretary for Science and Technology in the Department of
Commerce with respect to industrial innovation, including ETIP,
the Domestic Policy Review, our program for impacted industries,
our concept of the future cooperative technology program, and
finally, our attitudes towards the ownership of patents flowing
from government funded R&D.

You will appreciate that I have in this short time been able to
only brush the top of these items.

In closing, let me say that my statements have represented in
many cases my own views and not that of the Department of Commerce
and should be treated accordingly.

RECEIVED March 14, 1979.

Productivity in Federally Funded R&D Programs

ALBERT T. MAASBERG

Contract Research, Development and Engineering, The Dow Chemical Company,
Building 566, Midland, MI 48640

What is gained from federally funded research and development
programs? The answer to this question is very important because
the government is financially supporting about 53% of the total
research and development being done in the country this year.
Like it or not, over one-half of the total effort is controlled by
the Federal government, and it is vital that it be productive so
that all of us who have helped pay for this as taxpayers can benefit from the results. Industry also has a large stake in this
total R&D picture since about 43% is supported by and about 68% is
performed by this sector.

This control through funding means that the government agencies decide to a considerable extent who is going to do the R&D -
the Government itself, academia, non-profit research institutes,
or industry. Each of these has its own ideas about and measure-
ments of productivity and these vary considerably from one to
another.

Some Governmental researchers feel that they have been suc-
cussful if their project is to be continued for another year.
There are some academic people who measure success by producing a
paper for publication. Some non-profit research institutes believe that a project has been successful if there are five people
working on it this year whereas there were only three last year.
Unfortunately, there are some industrial people who judge accomplishment only by the amount of fee obtained by doing some contract research and development work.

By productivity we in industry mean that the work done by the
Government in-house or on a contract or grant has resulted in adding something to the knowledge about a basic concept, to a process, to a product, or to a service, which can be utilized commercially. The factors of timing, size and profitability are very
important in commercialization considerations. Most industrial
research prople feel that the best way to improve productivity is
to take over and do all the Government research!

0-8412-0507-8/79/47-105-031$05.00/0

Experience

An outstanding example of successful utilization of the
information developed on a government contract with our company is
the Artificial Kidney program with the National Institutes of
Health. This was a development which was started on a completly
Dow basis and carried to the point where the risk of success and
the investment required were both too high to continue to support
the project entirely on Dow proprietary funds. On the basis of
reasonable terms, favorable to both sides, an agreement was worked
out to carry the program through the later development phase. The
Artificial Kidney is now in commercial production by Cordis-Dow
and some other producers. This is an example of a successful
project but it should be noted that there was a high proprietary
cost during the final commercialization stage and that the initial
production facility required a large private investment. During
this period there was a real need for protection.

Other programs which have this kind of government support
include the sodium-sulfur battery; and coal conversion, both gasification and liquefaction; all with the Department of Energy.
Other well known productive programs outside the chemical area
are the miniaturization of solid state electronics and the communications satellites, both with NASA.

Problems

We all understand that there are some basic constraints under
which the Government must operate when doing or supporting
research and development. The projects are usually designed to
solve a public (that is, a Government) problem, not often to
supply a public need. The market is usually the Government itself in the form of defense, space, health, and environmental
requirements. Thus, this is not normally an area of origin of
new products which will contribute to the growth of the Gross
National Product. Only fairly recently have commercial opportunities developed in the energy and materials areas.

To insure the supply of an item resulting from a research
and development project, the Government should protect itself
with the necessary patent and data rights and this is understandable and proper in everyone's mind. However, there are individuals in Congress and in the agencies who carry this protection idea
to such an extreme that there is no opportunity for profit to
industry possible from participating in a program. This constraint is a very real and substantial block to any incentive for
innovation.

Recently there has been a very definite trend, as far as
Government support is concerned, toward R&D which will be utilized
in regulatory matters. Because of the nature of this work, the
amount of commercially applicable fall-out appears to be
decreasing.

Before discussing briefly some of these government policies
which are affecting productivity, it would be helpful to our
understanding of some of the problems to discuss some examples of
things which have happened.

Some years ago there was a Government requirement for a thick
magnesium alloy plate. The quantity involved for the particular
Government use was small while the future commercial use for the
developmental material was estimated to be relatively large. The
work could have been done in a few months with a relatively small
amount of money. However, in order to contract for the application development and a supply of the plate, the Government insis-
ted that the Dow proprietary alloy composition be made public
which would entail loss of background trade secrets. Since there
was no way at the time to handle this information on a confiden-
tial basis, from a business point of view there was only one
course of action, to turn down the contract opportunity. As it
turned out, the Government spent considerably more money and time
than would have been otherwise necessary to start from the beginning and support the work necessary to obtain a lower quality
plate.

During the development of membranes for reverse osmosis for
the desalination of water, the Government initially required that
the rights to background patents had to be turned over to them in
order for them to support any part of the work by people active
in the field. As a result, nothing was done for a number of years.
Irreplacable research and development time was lost until a
reasonable agreement was worked out under a modified patent policy
which both served the public interest and encouraged knowledge-
able contractors to participate.

Conflict of interest can be a problem. However, to guard
against such conflict, some of the measures being taken are counter productive. For example, in some instances, contractors
who have expertise in the manufacture and use of certain hazardous
materials have been ineligible to bid on contracts studying the
possible hazards in the handling of such materials. As a result,
background knowledge and experience is not used and has to be
developed again. Thus Dow's experience and background in bis
(chloro methyl)ether, a possible contaminant formed in formalde-
hyde using industries, could not be utilized.

On the other hand, for work on specific waste stream pollution there is sometimes a requirement that the contractor be
directly involved in pollution with this material. In this case,
there are people whose productive business is primarily solving
these waste pollution problems who are then not able to offer
their expertise! Also, if the polluter in the second example is
not in conflict of interest, why is the material handler in the
first example in conflict of interest? These are confusing
contradictions!

Policies

Patent and data rights are very important factors for industrial companies in making decisions on commercialization. Apprec-
iation of this business consideration seems to be very difficult
for Government, both legislative and agency people. Dr. Nat C.
Robertson, formerly vice president of R&D for Air Products and
Chemicals, covered this matter very well when he pointed out back
in January 1977 that the Energy Research and Development Administration's consideration of mandatory licensing is a good example
of the Government's lack of understanding of what is required to
motivate industrial organizations to participate in the relatively
long-range, high-risk programs necessary for achieving the goal of
energy independence. Providing only payment for research performed,
even with the inclusion of fees, does not provide enough incentive
to insure industry participation.

ERDA, now the Department of Energy, has further developed the
use of waivers of title to foreground patents and this has been
helpful. For a company which has done work in the field, however,
this is not as important a factor in commercialization considerations as the mandatory licensing, especially of background patents.
The public interest would be best served if the contractor were
given sufficient time in which to supply the subject matter
covered by the background patent in sufficient quantity and at
reasonable prices to satisfy market needs. Patent waivers are
usually complicated and take a great deal of time to negotiate.

In September 1977, Congressman Ray Thornton indicated that
his House Subcommittee on Scientific Planning and Analysis had
agreed that there is a very genuine, direct relationship between
the health of our research and development efforts and the health
of the national economy. He further stated that having many
different patent policies in the various agencies of the Federal
Government has an inhibiting effect upon innovation and research.

An effort is being made to establish a uniform Federal system for the management, protection, and utilization of the results
of federally sponsored scientific and technological research and
development in H.R. 8596, dated July 28, 1977, introduced by
Mr. Thornton and thirteen others. This is a move in the right
direction to improve productivity, innovation and commercialization but further modifications are needed especially in the areas
of background inventions. The basic approach is good but the
resulting administrative regulations could be a concern. We
understand that action on the bill will be slow because of the
objections of Senator Gaylord Nelson and the Administration.

When we first became aware of the program on cooperative
agreements we were encouraged. The early discussion indicated
that a mechanism might be established which would enable a contractor to do a job for the Government with a minimum of technical
direction and regulation from the agency. This would be a way to
improve productivity.

Our experience certainly has been that the way to really get things done is to clearly state the problem then assign a competent, reliable individual or group to the task and expect them to come up with the solution working pretty much on their own.

As far as we can tell, the Federal Grant and Cooperative Agreement Act of 1977 enacted February 3, 1978 does not really speak to this problem but only separates Federal assistance relationships from Federal procurement relationships. (See the paper by Michael Michaelis in this volume.)

In this area of regulation, the President of The Dow Chemical Company, Paul F. Oreffice, has pointed out that the Dow regulatory costs were 186 million dollars in 1976. This was an increase of 27% over the 1975 costs and the increase seems to be continuing at the rate of 25 to 30% per year. The part of the picture which really hurts productivity is that 37% of the 186 million, or 69 million dollars, are considered excessive Federal regulatory costs.

Objective and Recommendations

Our purpose in preparing this paper and participating in this symposium is to increase productivity in federally funded R&D programs and resultant commercialization. We feel that as the result of pointing out the problems and describing examples, the situation will be better understood by everyone and that existing practices will be modified. There are two major points that we would like to make.

First, the Federal patent policy should be made uniform for all the Government agencies and the mandatory licensing requirements should be modified. Title to foreground patents should remain with the contractor in all cost-sharing contracts with strong march-in rights where the contractor fails to make the goods or service covered by the patents after a reasonable time, in reasonable quantity, and at a reasonable price or of a reasonable quality. Mandatory licensing of background patents and data should be required only after the contractor has failed to produce in the time required to develop a commercial position. The fruits of research take time to commercialize and protection is required during this period or not many innovations will be carried through to the commercial stage. Once a new development is ready for commercialization, some protection from competition is necessary to encourage industry to risk the investment of capital in the new venture. Most of the Government owned inventions and developments which are freely available for licensing are lying dormant because business people are unwilling to invest in a venture which can be copied immediately if the first party demonstrates the commercial utility.

Second, productivity can be improved by developing a procurement procedure which would enable the Federal agency to develop a clear statement of the problem then turn it over to the people

most qualified by knowledgeability, experience and ability to
solve the problem and expect it to be accomplished with a minimum
of interference and certainly without costly excessive regulation.
We realize that clearly stating the problem is difficult, particularly in areas like pharmaceutical and agricultural chemicals
where the ground rules keep changing, but discussions like this
should be helpful in explaining the need for improvement.

RECEIVED March 14, 1979.

APPROACHES

Cooperative Agreements: A Key to Accelerated Industrial Innovation

MICHAEL MICHAELIS

Arthur D. Little, Inc., 1735 Eye St., NW, Washington, DC 20006

Last May, President Carter charged the Department of Commerce with the responsibility for spearheading a multi-agency, cabinet-level, Domestic Policy Review of Industrial Innovation. By April, 1979, this Review is expected to present the President with highly focused policy options to assist him in forging a coherent strategy to influence the rate and direction of industrial innovation in the United States.

That strategy -- whatever its principal thrusts -- can, I believe, be greatly strengthened if it takes advantage of the recently-enacted Federal Grant and Cooperative Agreements Act of 1977. This Act provides a legislative framework for those new institutional arrangements between industry and government that are so urgently needed to spur industrial innovation. The need for such new arrangements emerges from several recent studies, including one[1] that we undertook for the Department of Commerce. We concluded that:

- "Policies for federal funding of civilian research and development should be formulated in the larger context of the complex process of industrial innovation."
- "Federally-funded civilian research and development is not sufficient -- by itself -- to bring about technological change in the private sector to any significant extent."

In an earlier report to the National Science Foundation,[2] on "Barriers to Innovation in Industry," we noted that recommendations for public policy changes, offered by industry, government, finance, and labor, included:

- Designation of a focal point in the Executive Branch of the Federal Government to coordinate public policies related to technological innovation.
- Clarification of public policy objectives for technological innovation, e.g., international trade, productivity, consumer satisfaction, job creation, increased industrial competition.
- Increasing effectiveness of public policies by targeting them to be industry-sector specific where necessary.

0-8412-0507-8/79/47-105-039$05.00/0

- Articulation and aggregation of market demand for
  products and services purchased with government funds,
  so as to create additional market "pull" (to complement
  technology "push") in those areas where private market
  forces are insufficient to sustain innovation.

For this audience, I do not need to dwell on the distressing
symptoms and statistics that bespeak the loss of American pre-
eminence in technological innovation. A few illustrative points
will suffice to jog -- and shock -- your memory:

- The U.S. Balance of Trade in manufactured goods is in a
  serious decline. The projected deficit in such goods
  in 1978 is anticipated to be some $18 billion as part
  of the total projected $44 billion deficit. By contrast,
  Japan enjoyed a $63 billion trade surplus last year in
  manufactured goods. We had a $3.6 billion trade deficit
  with Japan in 1977 -- in high technology goods alone.
- In real terms (constant 1972 dollars), R&D spending in
  the U.S. has been on a plateau of slightly under $30
  billion per year since 1965 (and much of it is for non-
  civil purposes).
- The U.S. Patent Office issued fewer patents to U.S.
  citizens in 1973 than in 1963, but issued more than double
  the number of patents to foreign citizens in the same
  period. According to the National Science Foundation, the
  U.S. share of initiating important industrial innovations
  declined from 80% in the mid-50's to 54% in the mid-60's.
  All signs are that it is still on the decline.
- Brookings Institution reports that growth in U.S. productivity has been cut by 20-25% in 1975 by environmental,
  safety, and health regulations. In the mid-50's, federal
  regulation had major responsibility in four areas --
  antitrust, financial institutions, transportation, and
  communication. In the mid-70's, some 83 federal agencies
  are regulating many aspects of the private sector. Com-
  plying with federal regulations cost Dow Chemical Company,
  for instance, $186 million in 1976, equivalent to 50%
  of Dow's after-tax profits. Most disturbingly, federal
  regulation produces less willingness to take high risks
  in commercializing new technology because of the
  uncertainty of the regulatory climate.
- SEC reports that underwritings for companies with a net
  worth of under $5 million declined from 418 in 1972 to
  four in 1975. Yet, it is the small innovative, high-
  technology company that has historically been so often
  the wellspring of industrial innovation.

I could go on with more dismal details. But my purpose in
citing the evidence for declining U.S. technological capability
and industrial innovation is merely to spur you into action,
specifically with regard to the recently enacted law which is a
"sleeper" in that it provides a remarkable opportunity for a turn-
around.

Numerous studies, all the way back to the White House
Civilian Technology Panel in the Kennedy Administration (which I
was privileged to serve as Executive Director) have called for
new "institutional arrangements" to stimulate industrial innovation. The general perception is that nothing has happened in
response to these calls, and that apathy continues.

This need not be. President Carter signed into law (on 2
February 1978) the Federal Grant and Cooperative Agreements Act
of 1977 (P.L. 95-224). Even though it was admittedly not its
principal intent, this Act -- I submit -- provides a legislative
framework for the long-sought new institutional arrangements to
spur innovation -- provided that industry presses this point in
the Office of Management and Budget (Executive Office of the
President).

OMB currently is developing guidelines for all federal
agencies to implement the new Act. Proposed guidance was published in the Federal Register on May 19, 1978, inviting comments
by June 20. Practically none were received from industry.

OMB is also required by the Act to study alternative means
of implementing federal assistance programs, provided for in the
Act. A plan for this study was published in the Federal Register
on 23 June 1978, calling for comments by 23 August. Some industrial response has been forthcoming, notably from the Industrial
Research Institute. IRI suggested that "additional emphasis be
placed on evaluation of opportunities provided by the Act for
improving the effectiveness of federal involvement in technological innovation." It also recommended that OMB (in pursuit of
its study under the terms of the Act) work closely with that
Interagency Team of the Domestic Policy Review of Industrial
Innovation which will be addressing federal procurement policy
issues that impact industrial innovation.

It has been -- and still is -- an unfortunate feature of our
political life that industry appears reticent to come forward
with practical suggestions for public policies designed to improve
the climate for risk-taking in the private sector -- a necessary
prerequisite for industrial innovation. To be sure, industrial
leaders point to:

- debilitating features of high capital gains taxes, and
  inadequate incentives for high-risk business investment;
- failure to enhance capital formation and thus help expand
  and modernize productive sectors;
- government over-regulation that stifles creativity in
  technological progress and that diminishes productivity;

and a host of other factors which can best be summarized by
"uncertainty" of future market dynamics, induced by increasing
government intervention. Valid though these points are, their
effectiveness in impacting public policy for innovation is weakened by being aimed at widely scattered federal targets, each
supported by powerful groups of vested interests.

In this context I am pointing at the Federal Grants and Cooperative Agreements Act as a unique and timely target which
deserves fullest and forceful attention by industry during this

year and next -- when plans for its implementation are being made.
It is a unique target in that it encompasses all federal agencies
and in that it provides an opportunity for significantly increasing the productivity of taxpayers' dollars by stimulating industrial innovation and thus improving our balance of trade, increasing employment, and reducing inflation. This is an opportunity
we dare not let pass.

Why do I consider this Act as the legislative framework for
the long-sought new institutional arrangement to spur industrial
innovation? I can do no better than to quote from the recent
report by the Office of Technology Assessment of the U.S. Congress
on "Applications of R&D in the Civil Sector: The Opportunity
provided by the Federal Grant and Cooperative Agreements Act of
1977," published on 20 June 1978[3] Its findings are summarized as
follows:

"Federal R&D designed to stimulate technological change
in areas like energy, housing, and law enforcement are
effective only if non-Federal users adopt the innovations
produced. Federal management of such R&D must therefore
differ from that appropriate where the Federal Government
is the end user, as in defense and space R&D.

"The recently enacted Federal Grant and Cooperative Agreement Act requires that in all transactions with non-Federal
(civil sector) parties, Federal agencies distinguish between
'procurement' -- buying something for the Federal Government's direct use -- and 'assistance' -- supporting or
stimulating a non-Federal activity in the public interest.
Transactions to support non-Federal R&D would generally be
for the purpose of assistance. Yet, currently, much non-
Federal R&D is funded through the Federal procurement process. The change required by P.L. 95-224 presents an
opportunity to develop management perspectives and practices
appropriate for cooperative Federal/non-Federal efforts to
stimulate technological innovation.

"To clarify Federal roles and responsibilities, the Act
established uniform criteria for grants, contracts, and
cooperative agreements. These uniform Government-wide
criteria have the effect of forcing Federal agencies to
declare clearly which relationship with non-Federal parties
is sought.

"If Federal agencies are to become effective agents of change
through support of R&D, they must involve those non-Federal
parties -- whether in the public or private sector -- who
have the incentive and capacity to go beyond the R&D stage
and develop technological innovations for widespread use and
public benefit. The cooperative agreement is a new legal
instrument appropriate for such involvement. As in a joint
business venture, Federal and non-Federal rights and obliga-
tions are negotiated in the process of reaching such agreements.

"The Act mandates the Office of Management and Budget to make a comprehensive study of Federal assistance relationships and report to Congress in 2 years (i.e., in early 1980). The study presents an important opportunity to develop the new perspectives and procedures appropriate for assisting technological innovation. Because the OMB study will largely determine how the Act is implemented, Congress required OMB to involve in the study a wide range of potentially affected parties, including the Congress itself. Such involvement is essential in order to realize the Act's potential -- which is still not widely recognized -- for applying science and technology to a broad range of problems confronting the Nation."

For the sake of precision, as the OTA report notes, it is useful at this point to offer two definitions. The term "technology" is used here to denote knowledge required for the production and delivery of goods and services. This definition encompasses both physical and social technologies. "Technological innovation" (or "Industrial innovation") refers here to the process by which knowledge is developed and transformed into marketable products, processes, and services. The innovation process includes the whole gamut of steps in the development, testing, financing, production, marketing, diffusion, and use of a technology in the commercial marketplace.

Since World War II, the great bulk of federal R&D funding has been devoted to national security and space exploration. The principal reason that government has been successful in fostering innovation and technological progress in these two areas rests on the fact that government was procuring not only R&D but was also buying and using the products of that R&D. It both pushed technology through R&D and it pulled technology through using it in accomplishing the nation's defense and space missions.

In the last two decades the government has sought increasingly to apply technology to the solution of social and economic problems. To this end, it has funded R&D in such diverse fields as energy, environment, health, housing, transportation, education, law enforcement, and manpower training. What government failed to realize until recently is that what worked for defense and space -- i.e., R&D funding -- does not necessarily work in these civil areas where government itself is generally neither the delivery system nor the end user. Instead, it is private industry, financial institutions, and the consumer who determine what risks to take in the utilization of R&D, i.e., in the commercialization of innovative products, processes, and services.

Our report on "Federal Funding of Civilian Research and Development," that I alluded to at the beginning of my talk, provides ample evidence that federal R&D funding alone is not sufficient to bring about industrial, technological innovation in the private sector. In large part this is due to the fact that federal officials do not possess detailed knowledge of non-federal users' needs. Yet, such intimate knowledge of users'

needs is recognized by entrepreneurs, and by scholarly studies of
the innovation process, to be an essential prerequisite for
successful commercialization of technology, i.e., of the fruits
of R&D.

Another handicap which besets federal officialdom is its
lack of understanding of the calculus of risk-taking in private
industry and finance, particularly under conditions of mounting
uncertainties often engendered by the changing climate of federal,
state, and local government regulations. It is difficult enough
to track and predict the course of any particular category of
regulation. It is well-nigh impossible to anticipate the outcome
of trade-offs between conflicting regulations -- an outcome more
often than not governed by political power plays.

For instance -- and without implying any value judgment of
my own -- many OSHA, EPA, and FDA regulations appear as anti-
competitive, putting them in direct conflict with FTC and Justice
Department efforts to promote competition.

EPA and Interior Department regulations on mining and burn-
ing of coal and on production of shale oil, for instance, run
directly counter to Energy Department programs to encourage the
use of coal and to develop domestic resources of liquid hydro-
carbon fuels.

All too often regulations mandate design and product or
process standards. This stifles the search for innovative solutions to social and economic problems. Make no mistake, I do not
challenge the worthiness of social objectives of government regulations. But this worthiness does not justify government closely
regulating every facet of private behavior. There is a real need
for industry and academia to participate with government in the
debates on regulation. William Baker of Bell Laboratories has
suggested that they work as equal partners in defining appropriate regulatory systems. One feature of such systems could be to
work towards <u>performance</u> standards -- improved safety, better
energy efficiency, reduced air pollution -- letting industry
reach these standards in its own way, insisting only that it
reach them.

To repeat, lack of detailed knowledge of non-federal users'
needs and of the calculus of risk-taking in the private sector on
the part of federal agencies has demonstrably led to technological
pathways being pursued that -- with hindsight -- were found not
to meet the desired objective. Two examples, taken at random
from a sadly long list of such failures were "Operation Break-
through" in housing and much-vaunted "people-mover" systems for
urban public transportation.

Recognizing these fundamental deficiencies, we can see how
the Federal Grant and Cooperative Agreements Act can provide the
legislative framework for new institutional arrangements between
the federal government and non-federal parties of all kinds in
pursuit of not only commercializing federally-funded R&D, but
also in spurring industrial innovation at large. It can provide
a government-wide, institutional means of broadening the scope
and concern of federal R&D program managers to the entire process

of technological innovation in the private sector, rather than
just the setting and meeting of technological goals.

The Act distinguishes between three basic relationships.
The first type is that of <u>procurement</u>. This mode is indeed the
currently prevailing one. Here the executive agency is ultimately
responsible for assuring performance. The agency therefore must
establish the specific requirements to be met, judge the accept-
ability of the product or service against those standards, monitor
the work, and be involved to the extent necessary to assure prompt
and satisfactory performance. It has the right unilaterally to
change the work and terminate it for default, if necessary. The
Act requires that only contracts be used for procurement relationships as hitherto.

The second type of relationship is an <u>assistance relation-
ship</u> where the federal agency has little or no need for involvement during the performance of the activity assisted. The
agency's responsibility lies in defining the scope of the work
and in such monitoring as may be necessary to assure that the work
is performed within the agreed-upon scope. It is the recipient
who ultimately is responsible for assuring performance and expend-
ing funds within this agreed-upon scope, as in a basic research
grant. The Act requires that a type of grant be used to reflect
this relationship.

The third type of relationship also is an <u>assistance relationship, but one in which the federal agency is substantially
involved during performance</u>. In this case, the responsibility
for assuring performance is <u>shared</u> by the agency and the recipient. Correspondingly, defining the performance roles of the
respective parties also is a <u>shared</u> responsibility. As in a
<u>joint venture</u> between two private parties the whole range of
factors affecting the venture and its outcomes is the subject of
negotiation. These include: performance responsibilities, cost
sharing and cost recoupment, data and patent rights, termination
rights and procedures, cost accounting, subcontracting, and
liability and indemnification. The Act requires that a type of
<u>cooperative agreement</u> be used to reflect these relationships.

The Act places no restrictions whatsoever upon candidates
for assistance awards. Thus profit-making organizations that
were previously excluded from many assistance awards are now
eligible for them. And as the OTA report notes, "In view of
their central role in technological change, they are clearly
important candidates." While there may be problems in giving
federal assistance to private firms, since if effective it would
give the firms at least a temporary competitive advantage and run
the risk of displacing private funds with public funds, the report
concludes that openly competitive assistance awards would minimize
these difficulties.

It is the third type of relationship -- the joint venture
mode if I may so call it -- that seems to me to be the most
promising for commercializing federally-funded R&D and for stimulating industrial innovation.

There have been precedents for such joint ventures -- albeit under wartime conditions and not in all respects identical to foreseeable assistance relationships in a civilian, peacetime economy. Nonetheless, they are instructive to recall, as James Brian Quinn did last year before the Industrial Research Institute:[4]

"One of the main thrusts of S&T policy should be on devising and experimenting with new institutional arrangements appropriate to our priority problems and future demands. These will doubtless require rethinking and reshaping new relationships between government and decentralized, private, research and technology groups. In our investigations we found that the catalytic cracking, synthetic rubber, and Bell Laboratories programs offered some fascinating insights and guidelines for these relationships. Key elements in these programs are outlined as 'vignettes' below. "● In the summer of 1941 the Petroleum Administration for War was established to coordinate the development of petroleum products for the World War II effort. In four years a massive cooperation between the government and the oil industry increased 100 octane output 1000-fold. The government's main role was to provide 'the direction, coordination, red tape slashing, and encouragement to accomplish the impossible.' Although all the technical work was performed by private industry the PAW set clear priorities, eliminated fuels with octanes above 100, curtailed alternative uses of benzene and other aromatics that would contribute to 100 octane quality, broke transportation bottlenecks, and established incentives to offset the industry's losses on its production of other petroleum products. These included losses from facilities conversion, mix changes, and specification changes.

PAW arranged firm commitments for the government to buy 100 octane for a period long enough for industry to justify the enormous investments it would make. PAW pressed the development of refinery processes 'not yet beyond the laboratory and pilot plant stages. . . . In order that the fullest cooperation of the industry might be possible -- without conflict of antitrust laws -- PAW obtained Department of Justice approval for joint research and for exchange between companies and individuals of information concerning processes, products, patents, experimental data and general knowledge.' The initial endeavor was coordination of available productive processes. Later, the cost of risky scale ups of known development approaches was undertaken by the government. Through these processes, hydrofluoric and sulfuric acid alkylation, hydrocatalytic reforming, and fluidized catalytic cracking were all accelerated. And processes emerged which could produce 100 octane gasoline at commercial prices.

"● Prior to World War II various U.S. companies had been working on synthetic rubber processes. But no urgency was foreseen because the government's view was 'with the largest fleet in the world raw rubber would be accessible in a crisis.' But Pearl Harbor eliminated access to some 95% of such supplies. A government agency, the Rubber Reserve Company, was set up to help finance and bring on line synthetic rubber capacity for the war effort. But the agency lacked sufficient political clout to aid the fledgling industry get needed controlled materials for plants. The few plants Reserve Rubber got built in its first year were miniscule in output relative to needs.

In 1942 President Roosevelt appointed the Baruch Committee to study needs and recommend action. The Baruch Committee set high priorities for the program and established a Rubber Director, Mr. Jeffers, with great powers. He contacted all corporations in the field, told them the government was to serve the industry, and the industry was to press for what it needed to meet specified war and essential civilian demands. He decided that each company would adapt their facilities to whatever rubber they could best handle. But they must guarantee the quality and volume of output. Because of the crisis situation, cost considerations were sacrificed for output, and processes in development and pilot plant stages were pressed into production. The government relieved a shortage of tank cars for butadiene by having these built under priority conditions. The Rubber Director's Office also arranged an antitrust accommodation with the Justice Department, and process information was shared, with royalties -- if any -- to be worked out later. Exxon made its patents available royalty free.

The Rubber Director set priorities: to concentrate on basic rubber not specialties, to break the bottleneck on butadiene, and to produce rubber at whatever cost. The government undertook most of the development risk. It financed and owned the plants built, but these were planned, constructed, and operated by private companies. Individual companies also continued to develop their own processes separately in some cases. Under priority pressure for both 100 octane and butadiene output from the same fuelstocks and facilities, the oil companies found a way to increase yields of both simultaneously. Within 18 months it became possible to produce rubber on the scale needed. On the recommendation of rubber manufacturers a choice was ultimately made to concentrate on Buna-S rubber, one of many early possibilities. Technical work was carried on by the companies involved. The government's role was largely one of coordination, risk reduction, breaking bottlenecks, setting priorities, and ensuring demand. By 1944, 51 plants had been built, rubber supply had caught up with demand, and the Office of the Rubber Director was soon dissolved.

"● The Bell Telephone Laboratories (studied in the late
1950's and early 1960's) represented another productive relationship between private and public interests. Bell Laboratories was largely financed by a fee -- generally about 1%
at the time of my study -- allowed by rate setting bodies in
customer billing structures. Bell Laboratories' product development and design activities were paid for by manufacturing; costs were recovered through sale of products.

Because of the scale of the program and its funding base,
Bell could take on long-term fundamental research programs
that others could not. However, since the research funds
would be disallowed if not spent fruitfully, Bell had to
demonstrate that the gain to its customers, in the long run,
significantly outweighed the program's costs. Since rates
were pegged to Return on Investment, the customer's bill
would decrease with every efficiency gain by the Bell System.
But the company could gain through creating new service
opportunities for growth, avoiding possible preemption of
communication technologies by others, and enhancing its regulatory climate by improving the quality and cost of its
services. To ensure that its programs were closely matched
to customer needs Bell Laboratories developed a complex of
information flows, planning processes, and budgetary reviews
that brought Advanced Systems, Operating Company, Western
Electric, and individual customer preferences into research,
development, and design processes. And to make sure these
and scientific demands reached individual researchers the
laboratories had the most carefully worked out goal communication process I had ever seen in an R&D setting."

What do these -- and similar examples -- suggest as effective guidelines about potential partnership relationships between
business and government in meeting our future demands for large-
scale commercial systems? With some slight modifications I
agree with Jim Quinn that the government seems most effective in
stimulating innovation when it:

(1) Creates or Guarantees an Initial Demand: 100 octane
gas, synthetic rubber, computers, and cargo aircraft provide good
examples. Once private industry can foresee such a demand it can
invest its own money, become familiar with the product and its
production characteristics, and begin to develop technical cadres
that could support it in the private market phase. Competition
for the early market achieves the multiple competing designs,
personal motivation, and problem solving incentives necessary for
innovation. Interest in commercialization introduces economic
considerations early in the R&D and design process -- and that is
critically important.

(2) Breaks Down Bottlenecks: Synthetic rubber and catalytic
cracking provide excellent examples of the government's capacities
to break down barriers of secrecy, antitrust, transportation or
investment bottlenecks when this is in the public interest. By

developing better data on aggregate resources and setting priorities for use of scarce resources, development times can be significantly shortened.

(3) <u>Aggregates Demand</u>: By standardizing aviation gas, assuring demand for synthetic rubber tires, or through other actions (such as <u>appropriately</u> formulated standards for sanitation, food contamination, commercial broadcast, or waste disposal) government can aggregate market structures, making it easier and less risky for private parties to innovate for or product responsibly in those markets. When -- and that should really read "if" -- properly formulated, today's environmental standards or public purchases (as through the highway trust fund) can do the same. (Note my earlier remarks on performance standards!)

(4) <u>Aggregates Resources</u>: The Bell Laboratories' concept of aggregating research monies to serve the large-scale needs of a diverse using sector has been paralleled in the development of EPRI for the electric utilities. This concept could be extended into other areas where a fragmented industry -- like coal or natural gas -- has large-scale system needs that its individual companies could not finance.

(5) <u>Extends Time Horizons</u>: The Bell Laboratories' financing example and other actions (like the setting of 27-1/2 mpg fleet mileage standards for 1985 autos) usefully extend the time horizon of both government and private groups. Unfortunately political pressures -- and our private sector reward systems -- too often do the opposite, compressing time horizons to the 2-4 year frame of the election cycle, or the similarly short-time module of corporate top management. But through longer-term goal setting -- in conjunction with industry, and through quasi public financing -- with industry-controlled technical development, government could actively stimulate innovation in priority areas.

(6) <u>Takes Unusual Risks</u>: By underwriting prototypes no one company could risk and forcing alternative technologies into being to decrease overall national risk, the government stimulated rapid advance in the state of the art of synthetic rubber, catalytic cracking, computers, and advanced communications systems. Once the characteristics of these systems were known, the risk for private industry to carry them further became significantly reduced. Similar risk reduction is possible today.

(7) <u>Provides Incentives</u>: When the government has provided adequate incentives -- through allowed profits, tax relief, depletion allowances, or other means -- it has tapped the nation's extraordinary technical-innovative capacities, both small and large scale. When these incentives are removed -- as they often have been through tax, price control, or regulatory action in recent years -- talent naturally flows to areas where it will be rewarded. One has only to look at the new venture investment figures mentioned earlier or at the effects of the over-regulated gas and railroad industries to see the consequences of removing incentives.

It is these kinds of government roles that need to be
explored more fully and that need to be adapted to each specific
assistance-type relationship, through cooperative agreements with
non-federal parties as provided for in the new Act. Most importantly, this exploration and adaptation must reflect a mutuality
of purpose and understanding between government and industry that
is finally embodied through negotiation in the "joint ventures"
of cooperative agreements.

It is along these lines that I believe industry should urge
OMB to proceed as it develops guidelines for all federal agencies
to implement the new Act. At the very least, these guidelines
should make it mandatory for all agencies to declare -- and sub-
stantiate -- which procurement or assistance mode it intends to
select for each of its specific programs that bears on commercialization of new technology and why it believes it to be the most
effective in bringing about commercialization.

We do know a good deal about what makes institutions innovative and, indeed, what it takes to bring to bear our technological resources on our most pressing social and economic problems.
I believe that, provided the new Act is implemented imaginatively
and flexibly, private industry will respond vigorously in cooperating with government to undertake joint ventures, and is thus
likely to shoulder more financial, relatively long-term, risks
associated with technological innovation in the civil sector --
including the cost of R&D which, after all, is generally only a
relatively small percentage of all the funds at stake in the whole
process of innovation. It may well be that federal funding of
R&D will thus become less essential -- in the civil sector -- than
it now appears to be to those who shape our National Science and
Technology Policies.

But we must have the political will -- both in industry and
in government -- to focus necessary efforts. We must unshackle
our latent capability to discover and to invent -- particularly
in those areas vital to our international commerce and to our
domestic economy. And we must modify our institutional arrangements between government and industry -- with substantive contributions made forcefully by industry itself -- to encourage innovation, using all we know about this process. The Federal Grants
and Cooperative Agreements Act of 1977 provides us with a unique
opportunity to begin and to accomplish these vital tasks.

LITERATURE CITED

1. Arthur D. Little, Inc. (Michael Michaelis), "Federal Funding of Civilian Research and Development," Report to Experimental Technology Incentives Program, U.S. Department of Commerce; Westview Press, Boulder, Colorado, 1976.

2. Arthur D. Little, Inc., and Industrial Research Institute Inc., "Barriers to Innovation in Industry: Opportunities for Public Policy Changes," Report to National Science Foundation, 1973; available from National Technical Information Service, Washington, D.C., Reference Nos. PB 229899/AS (Executive Summary), PB 229898/AS (Full Report).

3. Office of Technology Assessment, U.S. Congress, "Applications of R&D in the Civil Sector: The Opportunity provided by the Federal Grant and Cooperative Agreements Act of 1977," 1978.

4. Quinn, Prof. James Brian, The Amos Tuck School of Business Administration, Dartmouth College, Hanover, N.H. "National Policies for Science and Technology: Managing Science and Technology for Future Growth." Speech before Industrial Research Institute (New York, N.Y.) at Greenbriar, West Virginia, May 31, 1977.

RECEIVED March 14, 1979.

The Federal Role in Industrial Energy Conservation Technology

DOUGLAS G. HARVEY

Office of Industrial Programs, Conservation and Solar Applications,
U.S. Department of Energy, 20 Massachusetts Ave., NW, Washington, DC 20001

The potential for energy conservation in the industrial sector is quite high. The sector consumes an estimated 37 percent of all the energy used in the United States. Furthermore, due to the industrial complex having been developed during a period of abundant and low cost fuels, most industrial processes are relatively energy inefficient. This conservation potential will grow over the next several years as industrial energy consumption is expected to increase by 50 percent, to 45 percent of the nation's total usage by 1985.

As stated, the energy efficiency of industrial processes is generally quite low. In some direct heating applications, efficiencies are as low as 10 to 15 percent. Even the more efficient processes such as steelmaking are only about 30 percent efficient. While it is not possible to achieve 100 percent efficient processes, it has been estimated that 30 to 50 percent of industrial energy could be saved with universal application of existing, emerging and advanced conservation technologies. Such an achievement could save 10 to 20 percent of the total U.S. energy consumption.

The existing capital stock in industry is estimated to have a present value of 750 billion. Clearly, reconstruction of these existing plants to utilize today's best available conservation technologies is not economically feasible. Selective retrofitting of the most promising current technologies is, however, practical and in the longer term, with increasing energy prices, industry is likely to develop and adopt more energy-efficient technologies as present process equipment ages to obsolescence and is retired. The major issue now becomes whether the rate of improvement of industrial energy efficiency is sufficient to meet national energy goals.

Industry traditionally waits for cost increases to become quite severe before adopting economically oriented countermeasures. Their initial reaction to cost increases is to simply pass them through to the ultimate consumer. In this instance, where fuel costs generally comprise a small percentage of the cost

of energy intensive goods, these costs could escalate quite dramatically before industry adoption of economically oriented countermeasures. This mechanism, coupled with the fact that the
economic and technical feasibility of many important conservation
technologies is yet to be proven, constrains industrial energy
conservation.

A federal program targeted at mitigating these economic and
technical risks, and designed to accelerate the implementation of
industrial options for utilizing existent but inadequately employed technologies and for energy efficiency improvements has
been established. This U.S. Department of Energy Industrial
Energy Conservation program focuses on:

- Existing but underutilized technologies for which a Federal
  action can be identified to stimulate implementation
- New technologies developed by Research, Development and
  Demonstration (RD&D) to provide energy conservation advancements with proven technical and economic feasibility
- Incentives such as tax credits to provide economic stimulus
  for industrial actions in the national interest
- A market-oriented commercialization effort to ensure accelerated technology transfer focused on specific industrial and
  end users and maximum implementation of these technologies.

Industry has and will continue to conserve energy on its own
and several factors indicate that industry will move spontaneously
to save energy: an historical trend over the past 20 years aver-
aging about 1 percent improvement annually; an acceleration of
this trend since 1972 resulting from sharply increased energy
prices and initiations of voluntary energy efficiency targets for
1980; and future incentives for conservation in the form of higher
energy costs, possibly a set of energy-use taxes and energy conservation tax credits in the National Energy Act.

The ability to maintain these trends in conservation by private industry alone may be limited. According to industry reports,
the recent accelerated improvements have been achieved through
housekeeping and low investment retrofits for which the potential
is now somewhat exhausted. Further results are seen as depending
principally on larger capital investments, and the continued trend
of conservation acceleration by the private sector on its own is
considered unlikely.

There are over 310,000 manufacturing concerns in the United
States and each has unique characteristics -- consumption, technology base, financial capability, degree of innovativeness and
investment decision making. It is this sector which must be
addressed if significant energy savings are to be achieved via
conservation. The Industrial Energy Conservation Program of DOE
focuses on processes applicable across all of industry (a horizontal thrust) and processes of the most energy-intensive industries
(a vertical thrust) to increase the energy efficiency and to substitute more abundant fuels for scarce natural gas and oil.

A 13.5 percent reduction in industrial energy per unit of

production output is possible in the 1972-1980 period, based on
technological feasibility and economic practicability. The program seeks, therefore, to remove the technological and economic
barriers to achieving the total potential savings. More specifically, the estimated impact of the program will be 1.33 quads in
1985 and 3.8 quads in 2000 for RD&D activities alone. The specific savings that will result from the new thrust to stimulate the
application of the existing but underutilized technologies and
those resulting from the Energy Policy and Conservation Act effort
have not been specifically determined, but are expected to be
extremely significant.

Objectives

With this background in mind, it is now possible to state
succinctly the Federal objectives regarding energy conservation
in the industrial sector.
- Achieve maximum penetration of existing and new energy conservation technologies in as short a period as possible
- Substitute, where possible, inexhaustible fuels for scarce
  fuels
- Minimize the energy loss embodied in waste streams of all
  types (discarded products, materials, and energies).

The Federal Role

The Federal role in research, development and demonstrations
is not clearly understood at first glance. One thinks of large
industries with billions of dollars and large research staffs and
it is not immediately understandable why there should be any
Federal funding of research, development and demonstration for
saving energy in industry. Many immediately conclude that industry will do the necessary RD&D, and that Federal monies could be
more effectively distributed elsewhere.

Industry will, of course, achieve significant energy savings
on its own and has historically averaged a little over one percent
per year improvement in annual energy savings. It is important to
examine more closely the industrial capital and R&D investment
decision processes to see what industry will not do without Federal stimulus.

Major industrial capital investment decisions are strongly
influenced by factors beyond those of simple profit maximization,
although rate of return is the single most important element in
a capital investment decision. A company will not risk shutdown
or less of market position for the sake of small gains in expected
profit. Investments which favor growth such as new product development are normally preferred over those which lower operating
costs even if both offer the same opportunity to generate profits.

Advanced energy conservation technologies are usually considered high-risk projects by industry since they are normally

unproven in industrial environments. Process changes directed toward energy savings usually affect other process parameters as well. Every change in an industrial process relates to changes in risk of slowdown or failure which may far outweigh the energy and cost savings to be gained.

Capital investment budgets of the industrial sector are allocated by widely varying priorities which are generally grouped as "mandatory" and discretionary." Energy conservation investments can be in either category although they are most often placed in the discretionary group unless they relate to continued energy supply or survival. The cost of energy constitutes a relatively small part of product costs in the energy-intensive commodity industries - as shown in Table 1 - and energy conservation investments are generally considered only after investments are complete for product or market development, OSHA and EPA requirements and capacity improvements. Energy conservation investments, as confirmed in a recent survey of capital budgeting practices, are treated in much the same manner as other investments but often requiring a much higher return on investment. Industrial decisions are made with management judgement applied after some form of quantitative analysis.

Table I.

Relative Energy Costs, All Manufacturers 1975

(BUREAU OF THE CENSUS, U.S. DEPARTMENT OF COMMERCE, ANNUAL SURVEY OF MANUFACTURERS, 1975)		
COST CATEGORY	COST—$ BILLIONS	FRACTION OF VALUE OF SHIPMENTS
PURCHASED ENERGY	23.19	2.28%
WAGES AND SALARIES	209.96	20.63%
MATERIALS	558.52	54.87%
OTHER COST AND PROFIT	226.18	22.22%
VALUE OF SHIPMENTS	$1,017.85	100.00%

Of the energy conservation options being considered, industry will more likely pursue those involving low to moderate technical and economic risk and high return on investment and those relating to continued energy supply.

A recent survey of corporate, research and development spending of 600 U.S. companies (Business Week, June 27, 1977) provides some significant insights as to which industries are dominant in overall R&D. Table 2 displays some of these data.

Table II.

Industrial Sector R&D Expenditures

INDUSTRY	TOTAL R&D ($ MILLIONS)	TOTAL R&D (% OF PROFIT)	% SHARE OF R&D BY DOMINANT CO'S	Dominant Companies
CHEMICAL	1438.8	39.7	45	DOW, DUPONT, MONSANTO
FOOD PROCESSING	301.9	12.4	35	CPC INT'L., GENERAL FOODS, GENERAL MILLS, QUAKER OATS
METALS & MINING	157.4	25.0	30	ALCOA
ALUMINUM ONLY	83.2	30.0	57	ALCOA
PAPER	111.9	12.1	59	INT'L. PAPER, KIMBERLY-CLARK, SCOTT
STEEL	124.4	17.6	77	U.S.S., BETHLEHEM
TEXTILES	23.9	10.1	42	BURLINGTON, FIELDCREST

The majority of R&D in the energy-intensive industries...as shown above...is by the chemical industry and the least is by the textile industry. The two right-hand columns are the most interesting, however, since they reflect the dominant companies in R&D expenditures. Two steel companies, for instance, account for nearly 80 percent of the total R&D expended in the steel industry. Similarly, two companies conduct 42 percent of the R&D for the widely fragmented textile industry.

It would appear, therefore, that the results of industrial energy conservation R&D by industry on its own would likely be held proprietary by a few dominant companies whereas Federally cost-shared RD&D results would be available to all industries. Government involvement, therefore, enables equitable dissemination of new energy conservation technology.

Targeting RD&D efforts by the Federal Government requires a closer analysis of the purpose of the private sector R&D expenditures. More specifically, identifying which industries are investing strongly in energy conservation on their own. A recent analysis of this type revealed the petroleum refining and chemical industries are directing significant R&D funding to energy conservation and the aluminum industry allocates a significant portion of R&D investment to energy efficiency improvements. These facts dictate that a greater degree of care be given the development of a Federal role in involvement with these industries and that additional analysis is required to avoid redundancy of effort. This does not necessarily mean, however, that there should be no Federal role.

Federally cost-shared research, development and demonstration

will increase the rate of private sector R&D expenditures and will
significantly accelerate the introduction of new higher risk,
higher potential programs with energy savings earlier in time and
at significantly less Federal cost than many of the supply options.
The Federal participation with key industries assures that the
RD&D is performed by the most competent talent available in the
nation's leading research oriented corporations and assures wide
dissemination of the RD&D results. The Federal leadership enables
development of cooperative inter-industry projects such as the
energy-integrated industrial park which may not be pursued by
industry alone. In addition, the Federal involvement will help
industry understand and more readily adapt to required regulations.
    The Government role in documenting and disseminating infor-
mation pertaining to industrial conservation is a traditional one
not dissimilar to the efforts of the Department of Commerce or the
Department of Agriculture over the years. The key difference,
however, between those efforts and the commercialization effort
of Industrial Energy Conservation is the selectivity and sharp
focusing of the industrial technologies. The technologies that
are energy conservative but underutilized by the private sector
are identified and analyzed to determine reasons for the lack of
market penetration and to ascertain whether or not the technology
needs Federal actions to stimulate its increased use. The Federal
role relative to existing technology is, therefore, primarily one
of analysis and dissemination of pertinent information to the
specific end-use industries. Some of these underutilized technologies will require proof of concept demonstrations to show the
merits, whereas tax credits or other incentives might be the answer to other instances.
    In summary, there is a role for the Federal Government to
participate in industrial energy conservation through the techniques of RD&D, tax incentives and industrial reporting programs.
The emphasis would be on identifying existing but underutilized
technologies, developing new energy-saving technologies which are
not redundant to the efforts of industry alone and to use every
available means to stimulate the early implementation of such
results.

Strategy

    The basic strategy is a program of cost-shared research, de-
velopment and demonstration of selected energy conservative technologies directed at processes that apply to a wide spectrum of
industries and processes which are specific to the most energy-
intensive industries. Together with a strong emphasis on engineering development and full-scale demonstration in industrial environments, significant program effort is placed on the identification
and transfer of existing but underutilized technologies, processes
and techniques to achieve energy conservation in the industrial
sector. Activities are selected on the basis of: high energy-

savings potential, acceleration of implementation, nonredundancy with efforts of private industry, the degree to which benefits accrue to fragmented industry without research funds, and the degree and appropriateness of cost sharing. Candidate projects are selected based on extensive analysis or risk, cost and benefit.

The Industrial Energy Conservation Program project selection process perhaps gives the greatest insight as to the entire strategy behind the Federal role. In this process, there is a recognition that each industry has certain return on investment criteria which vary with risk. This is depicted by the "private sector" area in Figure 1.

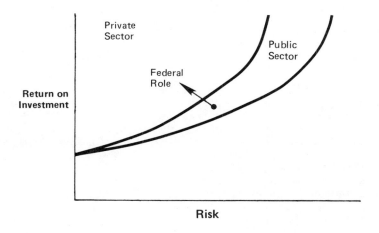

Figure 1

There is considered to be another area, consisting of somewhat higher risk or lower ROI programs suitable for Federal funding. This is shown in Figure 1 as the "public sector." The thrust of the Federal program is to perform sufficient RD&D on these programs to lower risk and increase ROI such that they become suitable candidate projects for private-sector sponsorship.

This philosophy of project selection for RD&D stimulation of existing but underutilized technologies is intended to have the effect shown in Figure 2. That is, energy benefits are achieved sooner than would be the case were the private sector left to its own initiative. The shaded area, therefore, shows the net benefit of the Federal program.

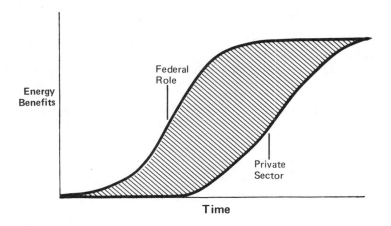

Figure 2

Selectivity and focus are key elements of the strategy. For the maximum impact to occur earliest in time, the projects must be characterized by high-energy savings, near term to realize the savings, nonredundancy to industry efforts, potentially acceptable to industry after the completion of Federal actions and environmentally acceptable and operationally safe. The existing but under-utilized technologies and practices are screened to include only those which have high conservation potential and for which a reasonable Federal role can be established ... incentives, assistance, demonstration, etc. The projects for Industrial Energy Conservation RD&D are primarily engineering developments of proven concepts and initiated as proposals from the private sector and academia. These are carefully screened and prioritized by a rigorous evaluation of cost, energy savings, clear establishment of a Federal role, competitive market penetration analysis and environmental impact. Only those not being pursued by the private sector on its own are considered and cost-shared contracts are initiated on the most promising activities. Every attempt is made to obtain cost-shared relationships with representative end-user companies for the demonstrations; with equipment manufacturers having the capacity to supply the market once the project is successful; and with representative trade associations. Having such a cost-shared program provides a vested interest of the elements of the private sector who will ultimately implement the technology. Having industrial "knowns" actually performing -- and contributing to -- the project has the effect of accelerating the market penetration once successful.

The focus of the RD&D is in two directions at the industrial targets: the energy-intensive generic technologies having wide

applications across the industrial specturm <u>and</u> the energy-intensive processes of the most energy-intensive industries of which a
few constitute a major portion of all industrial consumption.   By
this approach -- horizontal and vertical thrusts -- the relatively
low per unit energy-savings ideas with very large numbers of applications and the relatively high per unit energy-savings ideas
with a smaller number of applications, are both captured and
selected.

The degree of Federal effort required to affect suitable impact varies dependent on the technologies and range of applications.   The industry-specific technologies, for example, are expected to require fewer demonstrations than the generic technologies which have a widely diverse number of substantially different
applications.   Some efforts, such as waste-heat recovery, are expected to require second and third generation technologies which
build upon the results of preceding developments to cover all
applications.

Perhaps the best indication of the type of RD&D efforts that
are sponsored is by example.   The following gives descriptions
and results from two (of many) such efforts currently being pro-
secuted.

Coil Coating

Paint curing is a key manufacturing step in finishing many
kinds of metal products.   In the United States, paint curing consumes over 196 trillion Btu of energy annually, 95 percent of
which is supplied by natural gas.   Cutbacks in natural gas supplies can affect the many industries which use thermal curing for
paint systems, as well as those industries which depend upon these
products for manufacturing operations.   The coil coating industry
is one which is heavily dependent upon natural gas and their representative association, the National Coil Coaters Association
(NCCA), approached DOE concerning a solution to the natural gas
cutback problem.

One of their members, Roll Coater, Inc., agreed to join with
NCCA and DOE in a cost-shared demonstration project to utilize a
new technology that recovers a significant portion of the energy
normally lost in oven exhaust gases.   This new system, engineered
by B&K Machinery International, Ltd., makes use of newly developed
incinerators which burn the solvent vapors circulated to them from
the cured paint.   This recirculation reduces the ventilating air
burned along with some natural gas to supply thermal energy to the
oven for curing.   The final hot exhaust gases, which have some
unburned solvent fumes, are incinerated in a waste heat boiler to
generate steam which is subsequently used for heating the coil
cleaning tanks and for building heat.   This particular installa-
tion, which has been in operation since late 1977, has reduced
natural gas consumption by 45 to 65 percent for the oven and 65 to
85 percent with heat recovery from the waste heat boiler,

resulting in a total natural gas saving of 600 million cubic feet annually. Additionally, four other installations have also been made resulting in total savings of over 1.3 billion cubic feet of natural gas. The technology is continuing to penetrate the coil coating industry.

High Performance Slot Forge Furnace

Slot furnaces are in general use throughout the steel industry to heat steel for forging and are usually fired with light oil. Because many such furnaces are located in small businesses, it is difficult to obtain an accurate estimate of their total energy use. The best approximation available is that about 0.2 quad is consumed in slot forge furnaces annually.

The furnaces are usually inefficient, even in well-operated shops. Efficiencies of 5 to 10 percent are common. Under DOE sponsorship, Hague International of Portland, Maine, a manufacturer of furnace equipment, has developed a slot forge furnace that offers improvement through recuperation, excess air control, counterweighted slot-closing doors, and other conservation mechanisms. Recognizing the problems involved in marketing replacement furnaces, Hague International has also made available retrofit packages for existing furnaces. The data obtained thus far indicates that reductions in fuel usage of approximately 50 percent are available from retrofits, while savings of nearly 70 percent are achievable through furnace replacement. The Hague furnaces can be successfully operated with either light or residual oil, and data is being collected to establish whether it can utilize coal/oil slurries.

During FY 1979 a definitive and carefully monitored demonstration is being undertaken. A completely new furnace is being installed at Rockwell International's forging facility in Chicago, Illinois, and detailed records of fuel use and product throughput will be maintained. These data will be compared to records on conventional furnaces for a typical product mix. The purpose of the demonstration is to provide evidence and quantification of the potential savings for the forging industry through use of this type of furnace.

The potential savings from this technology are estimated to be approximately 0.13 quad annually. However, because of the fragmented nature of the forging industry and the fact that many operators will select partial retrofits which can cost as little as $20,000 against a total system cost in excess of $100,000, a savings of 0.07 quad per year of light oil seems a more realistic goal.

Commercialization

The previous paragraphs have discussed the origins of the bodies of technology -- existing and new -- and how these are selected

and dealt with. The critical task of moving these technologies
expeditiously into the marketplace and into the processes of industry is equally as important. The objectives of industrial
energy conservation cannot be achieved unless the private sector
itself puts the results to work.

The process of getting the technologies implemented by the
private sector -- called technology transfer, commercialization
marketing, outreach, etc. -- is a complex one for the industrial
sector. Unlike other sectors there is no broad readily understood market as in transportation or in residential/commercial.
The industrial market is highly diverse with each industry having
very different requirements, capital conditions, asset turnover
rates and differing degrees of innovativeness. Therefore, it is
not effective to broadcast the particulars of a given technology
to industry in general since most of those reached by such methods will not be concerned with that specific technology.

Each particular market must be analyzed to assess its particular needs, timing and other characteristics. The Federal
action that is effective with one industry is not necessarily
effective with another. The planning of commercialization starts
with the beginning of the project and the market potential is a
key factor in the project selection process. The planning of
commercialization that starts with the beginning of each project
is inclusive of the commercialization actions required during
development and, ultimately, to implement the project.

Commercialization of a technology or practice includes numerous potential elements depending on the individual situations.
In some cases, it is sufficient to transfer the related information to the specific industries who, upon seeing the economic
benefits and proven nature of the concept, will readily implement
it at an acceptable rate. Other industries might require more
tangible evidence of success and may want to see the demonstration
unit in operation and, in some instances, incentives might be
required to stimulate the industrial acceptance of new or existing
concepts. Tax credits for cogeneration and energy conservation
equipment in the National Energy Act, for instance, will accelerate many such technologies. The results of the industrial energy
conservation program will be closely monitored to establish a
measure of its impact and to identify needed improvements in the
commercialization process. The current industrial reporting program --- direct reporting and reporting through trade associations ... provides a ready vehicle for assessing overall program
impact. The specific market penetrations of individual projects
will be tracted to get a more specific indication of the program
effectiveness.

Summary

In summary, the strategy of the Industrial Energy Conservation
program is to select the most energy-conservative techniques that

exist today; develop new technologies that industry (for various reasons) will not do on its own; effectively transfer the technologies to the private sector; and stimulate the rapid penetration by the usual marketing practices ... documentation, seminars, films, television spots, trade shows, etc... and, where effective, Federal incentives as appropriate.

RECEIVED March 14, 1979.

NASA Technology Utilization Program

LEONARD A. AULT

Technology Transfer Division, NASA, 600 Independence Ave., SW, Washington, DC 20546

As some of you may already know, NASA has been operationally involved for some time in an active and aggressive effort to stimulate commercialization and the use of aerospace-developed technology. Some sixteen years ago, NASA established its Technology Utilization Program for this express purpose, and over this period NASA has learned a great deal about the process for the transfer of government-generated technology to the commercial marketplace.

I would like to briefly describe the nature of the operational transfer mechanisms embodied in that program, and relate to you some of our experiences in technology transfer and the results achieved since the program's inception.

The NASA TU Program was established in 1962 in response to a Congressional mandate provided in the National Aeronautics and Space Act of 1958. In drafting this enabling legislation, Congress took due note of the potential value of new technological advancements required to meet this nations's R&D objectives in space exploration. A provision of the Space Act required that NASA "provide for the widest practicable and appropriate dissemination of information concerning its activities and the results thereof." Since its establishment, the NASA TU Program has evolved an array of technology transfer mechanisms which range from technical information systems to adaptive engineering programs.

Publications Program

An essential first step in promoting broader utilization of NASA technology is letting potential users know just what NASA-developed information and technologies are available. This is accomplished by means of a series of publications.

Under the provisions of the National Aeronautics and Space Act, NASA contractors are required to furnish written reports "containing full and complete technical information concerning any invention, discovery, improvement or innovation" which may be developed in the course of work for NASA. These reports provide input to NASA's principal technology utilization publication, the Tech Brief Journal.

Issued quarterly, Tech Briefs is a current-awareness medium and a problem-solving tool for its industrial subscribers. Each issue contains information on more than 100 newly-developed processes, advances in basic and applied research, innovation concepts, improvements in shop and laboratory techniques, and new sources of technical data and computer programs derived from the many and varied aerospace R&D activities.

A special feature of Tech Briefs is a section on "New Product Ideas," innovations stemming from NASA research that appear to have particular promise for commercial application. Interested firms can follow up by requesting a Technical Support Package, which provides more detailed information on the new product or process deemed worthy of commercialization.

The journal enjoys favorable acceptance among its many industrial readers; the list of subscribers now numbers more than 45,000, and it is continuing to grow now at a rate of over 15,000 new subscribers annually.

The process of spreading the word is additionally aided by a cooperative industrial trade press, which republishes Tech Brief information for expanded circulation. In 1977 innovations reported in Tech Briefs generated over 120,000 requests for additional technical information, concrete evidence that the publications program is playing an important part in inspiring broad secondary use of NASA technology.

Another technology utilization publication, the Patent Abstracts Bulletin, deals with NASA-patented inventions available for licensing, which number now more than 3,500.

NASA sometimes grants exclusive licenses to encourage early commercial development of aerospace technology, particularly in those cases where considerable private investment is required to bring the invention to the marketplace. Non-exclusive licenses are also granted, to promote competition and bring about wider use of NASA inventions.

NASA also publishes Computer Program Abstracts, an announcement bulletin which advises of aerospace-developed computer programs available for adaptation to industrial or civil use.

In addition to these regular publications, NASA publishes a variety of special publications--reports, technical handbooks, data compilations--to acquaint the non-aerospace user with NASA advances in various states-of-the-art. Examples include new developments in welding and soldering, lubricants and lubricating techniques, human factors engineering, and sterilization and decontamination.

Numerous examples of technology transfer brought about by NASA Tech Briefs and other TU publications have been documented as part of our continuing program evaluation and user follow-up activities.

In one such example, the construction of a building in Washington, DC, (Figure 1) was based on a money-saving method of preparing building specifications which derived from a NASA system designed to obtain quality as well as minimum cost construction of launch facilities, test centers and other structures.

NASA's Langley Research Center developed a novel approach to providing accurate, uniform cost-effective specifications which can be readily updated to incorporate new building technologies. Called SPECSINTACT, it is a computerized system accessible to all NASA centers involved in construction programs. The system contains a comprehensive catalog of master specifications applicable to many types of building construction.

SPECSINTACT now enables designers of any structure to call out relevant specifications from computer storage and modify them to fit the needs of the project at hand. Architects and engineers can save time by concentrating their efforts on needed modifications rather than developing all specifications

Figure 1

from scratch. The NASA SPECSINTACT system has been modified
and adopted by the American Institute of Architects in a new
version which they call MASTERSPEC. The AIA claims that while
MASTERSPEC does save time and money, its use also involves no
sacrifice in architectural design freedom--a vitally important
consideration of their member firms.

Dissemination Centers

To promote technology transfer within the nation's industrial complex, NASA operates a network of Industrial Applications Centers (IACs), whose job it is to provide information
retrieval services and technical assistance to industrial
clients. The network's principal resource is a vast storehouse of accumulated technical knowledge, computerized for
ready retrieval.

Through the IACs, industry has access to some 10 million
documents, one of the world's largest repositories of technical
data. Nearly two million of these documents are NASA reports
covering every field of aerospace activity. In addition, the
data bank includes the continually updated contents of many
scientific and technical journals, plus thousands of published
and unpublished reports compiled by industrial researchers and
by government agencies other than NASA. Each month another
50,000 documents are added to this wealth of technical information.

The IACs seek to broaden and expedite technology transfer
by helping industry find and apply information pertinent to a
company's projects or problems. The philosophy behind the IACs
is that it is wasteful to "reinvent the wheel," that there is
no need to duplicate research already accomplished and thoroughly
documented in the data bank. Therefore, taking advantage of
IAC services, individual business firms--large and small--save
time and profit from research and development already conducted
by others.

Seven in number, the IACs are located at university campuses
across the country, each serving a geographical concentration
of industry. The IACs also have off-site representatives
serving industrial clients in many major cities and their
surrounding areas. Additionally, there are technology coordi-
nators at six NASA field centers who perform the important
function of matching on-going NASA research and engineering
with industrial interests.

Staffed by scientists, engineers and computer retrieval specialists experienced in working with companies, the Centers provide three basic types of services. To an industrial firm contemplating a new research and development program or seeking to solve a problem, they offer "retrospective searches" in which they probe the data bank (Figure 2) for relevant literature and provide abstracts or full-text reports on subjects applicable to the company's needs. IACs also provide "current awareness" services which are tailored periodic reports designed to keep a company's executives or engineers abreast of the latest developments in their fields with a minimal investment of time. Additionally, IAC applications engineers offer highly skilled technical and interpretive assistance in applying the technical information retrieved from the data bank to a company's best advantage. The IAC's charge nominal fees for their various services based on a value-added pricing policy.

A related service to industry is provided by NASA's Computer Software Management and Information Center (COSMIC) at the University of Georgia. COSMIC collects, screens and stores computer programs developed by NASA and other government agencies. Adaptable to secondary use by industry, government or other organizations, these programs perform such tasks as structural analysis, electronic circuit design, chemical analysis, design of fluid systems, determination of building energy requirements and a variety of other functions. COSMIC maintains a library of some 1600 computer programs, which are available to users at a fraction of their original cost.

Several brief examples of technology transfer made possible by NASA Industrial Applications Centers and COSMIC will underscore the value which these program activities add in bringing about beneficial change in U.S. industry.

In the first example, NASA heat pipe technology, used routinely for cooling spacecraft electronic equipment, was provided by the NASA Industrial Application Center at the University of New Mexico to Alaska Pipeline Service Company, the industrial consortium responsible for building and operating the Alaska pipeline. The upright supports of the pipeline shown in Figure 3, are heat pipes which keep the arctic ground frozen year-round, thus guarding against pipeline rupture by surface dislocations caused by seasonal freezing and thawing. As a result, NASA heat pipe technology plays a part in protecting the Alaskan environment from possible pipeline oil spills.

Figure 2

Figure 3

A structural analysis computer program called NASTRAN has been made available by COSMIC to a wide variety of industrial firms who have applied it to an equally wide variety of uses.

One such use was made by General Motors in the structural design of its Cadillac Seville. The use of NASTRAN improved the car's ride quality within weight limits and saved considerable development time. GM's successful application of NASTRAN to automotive structural design has since inspired the company to extend computer analysis to the entire GM line.

Another use of NASTRAN was made by PPG Industries, one of the largest U.S. manufacturers of flat glass. PPG designed and fabricated the frontal structure of a subway station in Toronto, Canada which is entirely made of glass. Transparent glass "fins" replace conventional metal support members used to provide support for wind resistance. At its glass research center near Pittsburgh, PPG Industries used NASTRAN, to analyze the stability of these all-glass structures under wind and load-bearing conditions.

Applications Engineering Projects

The information dissemination programs which I have just discussed are aimed primarily at the private sector. However, in the public sector, we have a different situation. Here our efforts are directed to demonstrating that aerospace technology can be useful in solving recognized public oriented problems in areas such as health, transportation, public safety, environment and so on. Since the primary beneficiaries of these projects are basically the public-at-large and not private industry, we work with other federal agencies on a cooperative basis.

For example, the Environmental Protection Agency (EPA) needed a small, portable device for monitoring water quality, to be deployed either from small boats or helicopters, EPA asked NASA for assistance. NASA's Langley Research Center developed a system which incorporates several aerospace technologies, particularly microelectronics, for processing water samples and automatically transmitting the resulting data. Shown undergoing test in Figure 4, the Water Quality Package was demonstrated to EPA in 1978.

Figure 4

The Water Quality Package is an example of NASA's "applications engineering" effort in the Technology Utilization Program. Applications engineering is the use of NASA expertise to redesign or reengineer existing aerospace technology for the solution of problems specified by other federal agencies or public sector institutions.

Applications engineering projects originate in one of three ways. The example just described illustrates how a government agency may ask NASA directly for assistance in the solution of an important problem. A second way is for a technologist at one of NASA's field centers to perceive the possible solution of a public sector problem by adapting existing NASA technology to meet that need. His proposal is then reviewed by the HQ Technology Utilization Office for technical feasibility, cost and other considerations. Project approval usually results in a cooperative, cost-sharing effort between NASA and the user agency. A project normally includes design, development, evaluation and field testing of prototype hardware to meet user agency specifications.

The third way an applications project may originate represents an innovative concept used by NASA to transfer technology to solve important public sector problems. The key elements are Application Teams consisting of several scientists and engineers who represent different disciplines. Located at research institutes and universities, these teams contact public sector agencies, medical institutions and trade or professional organizations to learn what significant problems might be susceptible to solution through application of aerospace technology. Having identified a problem, they then contact appropriate individuals at NASA field centers to determine what existing technologies might be adapted or applied to the problem at hand. Matching technology to need, the teams often conduct technology demonstrations as a first step toward bringing about commercialization or institutional acceptance of the technology transfer. Existing NASA application teams currently concentrate their efforts in the fields of medicine, public safety, transportation, and in improving manufacturing processes for increased industrial productivity.

The following examples serve to illustrate the various application engineering projects undertaken recently in cooperation with other federal agencies.

One of the operational requirements of the U.S. Coast Guard is to acquire a capability for a quick response to harbor and open sea fires. To meet this requirement NASA has developed a lightweight firefighting module transportable by helicopter to a number of existing ships. This module is capable of pumping greater quantities of water faster and farther than any other currently available system. This unit is now in preliminary test stages to meet the Coast Guard requirements. Incidentally, I should add that the NASA technology came from our work on high speed rocket engine pumps at the Marshall Space Flight Center.

The final example of our applications engineering activity is a portable, hand-held X-ray instrument developed at the Goddard Space Flight Center. This device which is called a Lixiscope (an acronym for Low Intensity X-ray Imaging Scope) resulted from our work on X-ray and gamma-ray spectroscopic techniques for astrophysical and planetary observations.

The Lixiscope (Figure 5) is a relatively simple device which is powered by a pen-light battery and utilizes a small radioactive source to produce low intensity X-rays. The Lixiscope consists of a viewing screen which permits real-time scanning of objects which are place between the scope and the X-ray source. The X-ray source is contained in a small lead-lined metal cylinder, not much larger than a thimble, mounted on the end of an extendable rod. The object to be examined is placed between the source and the scope. The Lixiscope is then triggered, and the source is unshielded. Low intensity X-rays then pass through the object and are converted and amplified through several unique process stages and finally converted to visible light which is then projected on the viewing screen.

Potential applications of this device are to be found in medicine, dentistry and areas of industry. The most obvious promise of this unique unit is in medical or dental emergencies and other field use where a quick fluoroscopic examination is desired; such as, (1) examination of a football player's possible bone injury on the football field; (2) root canal analysis and possible monitoring of surgical procedures; and (3) industrial detection of welding defects or gas leaks in pipes.

NASA is working with several research institutes in the dental and medical field to clinically evaluate the Lixiscope. The commercialization potential of this device is high. We say this only because many medical and other manufacturing companies have inquired about the availability of the device. The Department of Defense has identified many potential applications, individual practitioners and veterinarians have inquired as to its availability. The experience and information gained from the clinical field evaluations mentioned earlier will be invaluable to potential manufacturers in the commercialization of this technological "Spin-Off" from NASA.

This concludes my brief overview of NASA's Technology Utilization Program. As I stated earlier, much has been learned by NASA about the technology transfer process -- learned as a result of "doing" rather than"study." The technology transfer process is a complex one -- complicated on one hand by the sheer volume and rate of technological advance in recent years, and, in the case of NASA, complicated on the other hand by its goal to apply technologies across inter-organizational boundaries to problems or situations different from those for which the technology was originally intended.

Before I conclude my remarks, let me leave with you a conceptual framework for technology transfer for your future consideration ... a framework that characterizes the many and varied transfer mechanisms employed in NASA's Technology Utilization Program.

We have learned that there are essentially three basic types of transfer mechanisms: (1) information dissemination; (2) personal interaction; and, (3) applications engineering... which we euphemistically call paper, people and product mechanisms, respectively. Figure 6 illustrates these mechanisms as continuum of activities through which technology flows from their points of origin to tangible application in the user community. This continuum of transfer mechanisms represents a series of iterative steps designed to optimize the flow or transfer of technology from left to right with each step having its own added value characteristics. The first of such steps could be called an "awareness" phase (e.g., announcement of technology availability through, say, the NASA Tech Brief Journal); the second step is providing greater detail through, say, a Technical Support Package; then followed if necessary by personal contact by phone or visit between the potential user and the NASA innovator.

Figure 5

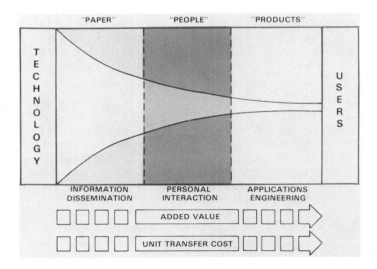

Figure 6.   Technology transfer

Likewise, computerized searching of the technical information data base to meet needs or problems specified by an industrial client, as performed by a NASA Industrial Application for example, is a paper/people mechanism. That is, the interaction between the search strategist and the user further aids in focusing available information on the problem at hand. This focusing process is illustrated by the funnel-shaped region in the figure as we move to the right through the transfer mechanism continuum.

Quite naturally, as each iterative step in the transfer process is brought into play, value is added to the process and a concomitant increase in unit cost per transaction is required. In NASA, for any given potential transfer opportunity or transaction, we always begin the process at the extreme left and move to the right only as far as necessary to effect a tangible use or transfer result. This approach has typically been called the "technology push" process. "Market pull," on the other hand, can be defined as the remaining distance between the point where NASA ceases to add value and technology and where adoption by the user begins.

NASA's experience has indicated that while most of its transfer mechanisms operate on the basis of "technology push" rather than "market pull," the need for the latter is often essential to achieving successful transfer. We have successfully demonstrated this fact, we feel, in working with other Federal agencies in effecting solutions to problems in the public sector. And indeed, similar results have routinely occurred in the private sector with industrial companies who are aware of their technological needs, have the ability to articulate these needs in a readily understandable form, and are willing-to-pay for technologies which are applicable to their needs.

In addition, we have learned that information dissemination mechanisms, although important to the process, are usually not sufficient in themselves in effecting successful transfer. Early recognition of this fact prompted NASA to evolve this broad array of transfer mechanisms which could be employed as necessary to achieve meaningful end results. This active and dynamic approach to technology transfer, while difficult to control programmatically, has substantially increased the Agency's effectiveness over the years in moving its technology out of the laboratory into the industrial and commercial marketplace.

RECEIVED March 14, 1979.

Patents and Technology Transfer

WILLIAM O. QUESENBERRY

Patent Counsel for the Navy, Dept. of the Navy, Office of Naval Research,
Arlington, VA 22217

World War II disturbed the tranquility of science and technology. Prewar Federal sponsorship of about forty million dollars a year in research and development has exploded to where three times that amount in tax dollars is spent each day in the current fiscal year. In doing so, the Federal Government supplies the bulk of research and development funds to the U.S. economy. To bring the matter a little closer to home, half of this Federal outlay is represented by the eleven-plus billion dollars allocated to defense research and development.

Research and development sponsorship has produced and accumulated, particularly within the Military Departments, a vast amount of new technology. This technology has served the defense establishment well in its mission to develop and acquire the weapons systems and materiel necessary for the defense of the nation. Large defense and aerospace contractors have transferred aircraft, air control and safety, computer and similar technology from the military sector to the civilian sector. Beyond that, there seems to have been little additional return on investment to the taxpayer in terms of use of this accumulated technology by private industry in its pursuit of the civilian market.

In history's first Presidential Message on Science and Technology to the Congress in 1972, it was acknowledged that an asset unused is an asset wasted. The President stressed the need to apply Government-generated technology to solving the nation's social and economic problems and bolstering American leadership in trade competition. This seemed to be the signal for executive agencies to organize and support effective programs for transferring mission-serving technology to wider use in the private sector. The flurry of awareness and organization for technology transfer is now quite apparent in most agencies. Whether or not this effort will be successful in attracting entrepreneurs to Government technology remains to be seen.

Scope of Military-Sponsored Technology

Looking back in time, we might ask ourselves why it is that
a storehouse of some 200 to 300 billion dollars worth of technology, free for the taking, has not been snapped up by private
industry. Insofar as defense technology is concerned, the
reaction by many to this question might well be "Who wants to
commercialize torpedoes, guided missiles, and tanks?" This is
a misconception of the true makeup of technology generated by the
Military Departments. Indeed, weapons and weapons systems are
what it is all about. However, for the most part these are
comprised of components and improvements having other applications. Also research and development to equip and care for
military personnel and facilities generates new technology having
widespread nonmilitary application. As a result, probably no
more than about one out of five inventions in the Navy's portfolio of some 10,000 patents are devices applicable solely to
military usage.

Government-financed research and development in furtherance
of the Navy's mission has produced technology in such fields as
medicine, chemistry, communications, transportation, energy,
environmental control, safety, construction and metallurgy, to
name a few. Illustrative of recent Navy patents in these fields
are inventions entitled: stand-up wheel chair, blood pressure
monitor, improved EKG contact, hospital patient monitoring
system, measurement of electrical impulses in the eye-brain
system for eye examination of very young children, mechanical
arm, antifouling paint for boats, gasoline additive, underwater
adhesive, microphone and headset for underwater swimmers, sail-
boats, solar panels, oil spill recovery, air pollution control,
noise suppressors, trash dump system, waste processing, desalination system, anti-air collision system, air passenger safety and
survival, anti-derailment sensor for trains, firefighting system,
etc.

Deterrents to Technology Transfer

Since military sponsorship of research and development does
in fact generate technology which has potential for civilian use,
there must be other reasons that it so seldom makes an appearance
on the commercial market. One reason might well be that few
entrepreneurs are aware of the military-sponsored technology that
is available for nongovernment use. Over the years, the Military
Departments have not seen it in their mission to imitate the
efforts of agencies such as the Department of Agriculture and the
National Aeronautics and Space Administration to aggressively
publicize to private industry and particularly small business
concerns what technical innovations are available. In addition,
normally a certain amount of technical information and assistance
should flow from creator to producer if effective and economical

commercial development is to occur. Here again, perspective as
to mission, funding, need-to-know, etc., has done little in past
years to encourage this ingredient of successful technology
transfer.

Thirdly, it must be appreciated that research that produces
new technology is but a small part of the cost of bringing that
technology to the market place. Far greater private risk capital
is needed at this point than has been expended by the Government
in its research and development phase. Authorities from the
business world estimate that for each dollar spent for inventive
activity, ten dollars is required for development of a working
model with commercial appeal and one hundred dollars to tool up,
manufacture, promote and distribute. The investment risk of
bringing untried inventions to the commercial market place must
be protected from coattail riding by would-be competitors or the
prudent businessman wants no part of the venture. This is especially true in the case of the smaller manufacturer who, having
developed and promoted a new item for the market, can subsequently be out-produced and under-priced by a larger competitor
with production economies of scale, extensive distribution
channels and no development investment to amortize.

The protection of private risk capital has been ignored in
the past by the Military Departments. Historically, the thousands of inventions in the military patent portfolio (which
represent two-thirds of the patents held by the entire Federal
Government) have been available only on a royalty-free nonexclusive basis--essentially public dedication. The poor record
of commercialization seems to give credence to the old adage
"that which is available to everyone is of little value to
anyone".

Transfer of technology generated under military research and
development programs will not come automatically or even easily.
At best, it represents a high risk, long lead time effort for
both the agencies and prospective users. Transfer of technology
must take place in the Military Departments in an environment of
increasing mission requirements and decreasing resources so
priorities of funds, manpower and objectives inevitably arise.
At the same time, obstacles of inertia, skepticism and concern
over investment return confront the private sector.

Invention Licensing

In the current effort by the Department of the Navy to carry
out the technology transfer mandate, the Naval Material Command
is moving forward in a positive manner in the areas of technology
analysis, publicity and technical assistance. In support of the
NAVMAT program, the Office of Naval Research, which has Navy-wide
responsibility for patent matters, has inaugurated a positive
licensing program for its portfolio of patentable technology.

This program began in 1976 when the Secretary of the Navy implemented Government-wide licensing regulations issued by the General Services Administration. The basic premise of the Navy's licensing program is to encourage the earliest possible use of Navy inventions by using the incentives of the patent system. Navy inventions are no longer considered dedicated to the public nor is a license granted or implied in a Navy invention outside of the framework of Navy licensing regulations.

Navy inventions covered by a U.S. patent or patent application, except those subject to security classification, are made available for licensing by the Office of Naval Research. Lists of available inventions are published in the Federal Register, the Official Gazette of the U.S. Patent and Trademark Office and through the technical publications of the National Technical Information Service of the Department of Commerce. At this stage, if an applicant is willing to commercialize an available invention on a nonexclusive basis, and shows the intent and capability to do so, the Navy will grant a nonexclusive license since this leaves the invention available for additional licenses to other interested parties and serves to promote competition in industry. This license is royalty-free and continues for the life of the patent as long as the licensee continues to make the benefits of the invention reasonably accessible to the public.

However, interest by the private sector in technology available only on a nonexclusive basis is limited. Therefore, to obtain commercial utilization of some inventions, it may be necessary to grant an exclusive license for a limited period of time as an incentive for the investment of risk capital. Accordingly, if an invention has been available for licensing for a period of six months with no qualified applicant for nonexclusive licensing and a prospective entrepreneur is interested only if protected by exclusivity, a limited exclusive license can be negotiated on terms and conditions most favorable to the public interest. In selecting an exclusive licensee, consideration given includes: his capabilities to further the technical and market development of the invention, his plan to undertake the development, the projected impact on competition, and the benefit to the Government and the public. An exclusive license gives the licensee the right to practice the invention for a period of time less than the remaining life of the patent. Normally this would be a period of five to seven years depending upon the nature of the technology. The concept is to allow one to three years (more in the case of commercialization requiring approval of the Food and Drug Administration or Environmental Protection Agency) for investment of funds and development of the invention for the market and a period of time at least long enough for the licensee to recoup his costs by exploitation of the invention. In exchange, the exclusive licensee agrees to invest a specified minimum amount of money and commit specified resources and effort

toward commercializing the invention and agrees to continue to
use his best efforts to practice the invention for the term of
the license.   Failure to live up to the agreed conditions may be
cause for revocation of the license by the Navy.

A royalty provision and/or other consideration flowing to
the Government is required in exclusive licenses, each case being
considered and negotiated on its own merits.   In all situations,
the commitment of risk capital and the benefit to the public
derived from commercial utilization is the prime objective of the
transfer of the technology.   However, in most instances a fair
royalty to the Government, payable in some cases after recoupment
by the licensee of his investment, is considered appropriate and
is normally acceptable to licensees.

A more aggressive promotional and licensing approach to the
transfer of Navy technology seems to have produced an encouraging
trend of interest by the private sector.   Licensing inquiries to
the Office of Naval Research jumped from 28 in Fiscal Year 1975
to 93 if Fiscal Year 1977, an increase of 230%.   As the result of
this interest, the Navy was able last year to get commitments to
commercialize eleven inventions on a nonexclusive basis.   More
importantly, in the first eighteen months since inauguration of
the Navy's policy of exclusive licensing, seven such licenses
have been granted with proposals to commercially develop five
other inventions now in the negotiation stage.   This represents
spin-off utilization of technology which would not come about
without the incentive of patent protection.   Also, for the most
part, interest seems to center around small business concerns who
find exclusive licensing an aid in protecting their entrance into
the market and an inducement in obtaining necessary financial
backing.

Case Studies

The advantage of patent licensing as a component of technology transfer is best illustrated by case study.   One example
would be a Navy laboratory-conceived invention in the art of
disinfecting.   The method involves an aqueous hypochlorite
solution which can be applied to sensitive surfaces such as skin
or clothing, as well as nonsensitive articles, since it self-
destructs after a short period of germicidal activity.   The
invention was patented by the Navy in 1973, but was never developed for actual use.   A small company organized by retired
medical and medical services personnel to develop and market
products in the field of disinfectants became aware of the patent
and interested in its development.   For two years, it was unsuccessful in getting financial support to cover required testing
for Food and Drug Administration and Environmental Protection
Agency approval and the development of an appropriate container-
applicator system because at the time Navy policy was to license
only on a nonexclusive basis.   Once a period of exclusivity to

protect investment became possible, an enthusiastic development
program began and financial support became a reality. The company is optimistic as to successful application of the technology
to various fields of industry.

In another example, Navy-sponsored research at a university
produced an invention in bone fracture healing through the use
of direct current from a portable power source. This invention
was patented in 1974. With the advent of the exclusive licensing
policy, a corporation engaged primarily in the research, development, manufacture and world-wide marketing of orthopedic devices
offered the Navy a plan of commercialization. This included
risk capital in excess of two million dollars to cover further
development and test work necessary to obtain approval of the
Food and Drug Administration to market the invention. Again, a
period of exclusivity in patented technology turned out to be
prerequisite to commitment from the private sector.

International Technology Transfer

Except for the National Aeronautics and Space Administration
and the former Atomic Energy Commission, Federal agencies have
generally ignored the foreign commercial potential of their
technology and relatively little foreign patenting has been done
by the U.S. Government. As a result, foreign manufacturers have
been able to exploit U.S. patented technology abroad and American
industry has had no patent protection under which to practice
Government technology outside the United States. Governments of
other industrialized countries have been less naive and traditionally protect significant inventions under the U.S. patent
system.

To protect U.S. technology abroad, the Naval Material
Command, with the assistance of the Navy patent staff of the
Office of Naval Research, has added a modest experimental international program to its effort. Two inventions, one in the
communications field and the other in the field of industrial
temperature control, have been selected and patent applications
filed in selected countries in Western Europe and in Japan. A
successful international technology transfer program supported
by foreign patent protection could aid in the protection of Navy
technology from exploitation by foreign interests, a more
favorable balance in import-export flow and access to important
foreign technology potentially useful to the Navy.

Summary

All Government bodies are charged with particular missions
and responsibilities. Those that provide for the national
defense or the improvement of the public welfare seek better
devices, systems and services directly needed to carry out their
governmental function. This is accomplished with the improvement

and advancement of technology in Government laboratories and
through contracts for research and development with the private
sector.   In the national interest, an objective of agencies
engaged in research programs must be to encourage widespread use
of the improved technology beyond just governmental use--to still
broader ends of national policy including promoting scientific
progress, the advancement of knowledge generally, and above all,
economic growth.

At the present time, the public is being taxed at an annual
rate of about twenty-four billion dollars for Government-
sponsored research and development.   The major portion of this
is directed toward national defense and space accomplishments.
However, the knowledge generated involves all branches of technology and is being largely underutilized at a time when the
economy needs all the help it can muster.   If it were channeled
to commercialization, in all probability the nation's economy
would be enhanced, new business enterprises would be organized
and the operations of existing business enterprises expanded,
with resulting increase in employment, improvement in the standard of living, increase in tax revenue, and improvement in choice
and price benefits to the consumer (including the Navy).   As the
real purchaser of research, the taxpaying consumer is entitled
to additional commercial benefits from his research and development tax dollar.

To this end, the Navy's technology transfer effort is
designed to combine active promotion and cooperative technical
assistance with a licensing program which uses the incentives of
the patent system as a catalyst for encouraging the transfer of
Navy technology into the stream of domestic and international
commerce.

RECEIVED March 14, 1979.

Commercialization of Technology Through the Federal Laboratory Consortium for Technology Transfer

CHARLES F. MILLER

Lawrence Livermore Laboratory, Box 808, L-790, Livermore, CA 94550

INTRODUCTION

The U. S. taxpayers have an enormous investment in Federally funded research and development (R&D). The latest report on federal laboratories (1) states that there are almost 800 R&D laboratories and centers located throughout the nation. These employed, in fiscal year 1977, more than 240,000 personnel with an operating budget of almost $8 billion. In addition, extramural research contracts amounted to over $7.5 billion. This investment in the federal R&D establishment can be viewed as a storehouse of ideas, hardware, facilities, equipment, processes, capabilities, experience, and individual expertise. These "technologies" may be of use to industry in the form of new products, product or process improvements, technical advice, or state-of-the-art information concerning on-going research projects.

A process of active technology transfer is widely viewed as a requirement to successfully and expeditiously transfer the results of federal R&D. Passive transfer systems do not provide timely awareness of potentially useful technologies and often fail to provide the user with necessary detailed information. Several federal agencies and laboratories do indeed support and conduct active technology transfer programs. Notable among these are NASA's Technology Utilization Program (2), the Department of Navy's "Technology Transfer Fact Sheet" and Navy Patents licensing program (3), and Oak Ridge National Laboratory's "Technology Utilization Bulletin" (4). In addition, these Federal organizations, along with many others, participate in and are members of the Federal Laboratory Consortium for Technology Transfer.

The Structure and Purpose of the Federal Laboratory Consortium for Technology Transfer

The concept of the Federal Laboratory Consortium for Technology Transfer began in 1971 with an informal network of Department of Defense (DoD) Laboratories. These labs held periodic meetings to exchange ideas on ways to disseminate DoD-developed technology to non-military users. In 1974, the major Federal R&D

laboratories and centers were invited to join the DoD Consortium
and the Federal Laboratory Consortium for Technology Transfer (FLC)
was established. Today, the FLC has grown to over 180 of the
largest Federal government research and development laboratories
and centers.

These laboratories and centers represent eleven Federal
agencies, including the Departments of Defense, Transportation,
Energy, Interior, and Commerce, NASA, EPA, and the FBI. Each FLC
member, or group of members, supports a Technology Transfer representative who, in addition to representing his or her own laboratory, maintains contact with other research institutions and other
Federal, private, and public agencies, thus forming a national
network of individuals dedicated to technology transfer. The
Division of Intergovernmental Science and Public Technology of the
National Science Foundation and the Naval Weapons Center, China
Lake, California, provide resources which make possible operation
of a Secretariat in support of Consortium activities.

Although there are many definitions of technology transfer,
within the FLC, it is generally described as the process by which
existing knowledge, facilities, hardware, or capabilities developed under Federal R&D funding are transferred to fulfill other
public or private needs. The Consortium serves as a forum for the
discussion of the principles and practices of technology transfer
and provides a communication network for the purposes of:

1. Facilitating the exchange of technical information, the
   diverse application of R&D results, and transfer of
   technology from the government laboratories toward the
   solution of existing problems and the avoidance of future
   problems in both the private and public sectors;
2. Encouraging the collection, compilation, and dissemination
   of information on existing technology transfer techniques
   and methodologies and experiences in their application;
3. Encouraging the development and implementation of technology transfer techniques and methodologies; and
4. Providing a baseline of experience for assisting decision
   makers in the development of national policy for technology transfer.

To accomplish these goals, FLC operation is aimed at eliminating or at least minimizing the effects of those barriers or constraints that may hamper technology transfer efforts of Federal
laboratories. The FLC emphasizes person-to-person communication
between the resource people (Federal or other R&D organizations)
and the users (state and local governments, educational institutions, and private industry). The core program activity is the
development of an organized information system and the involvement
of resource people and users in the problem definition and transfer process along with linking agents, or technology transfer
"brokers," to bridge the communication gap between researchers and
users.

The early efforts of FLC members were directed toward providing assistance to state and local governments. Assistance has
been provided in such areas as transportation, public safety,
economic development, energy, public works, environmental improvement, and community development. In these activities, continuous
efforts are made to involve the private sector to the greatest
extent possible. This "involvement" ranges from encouraging representatives of industry to provide technical assistance and advice
to local governments to helping create new products or to aggregate local government markets through field test and demonstration
of products or preparing standard procurement specifications (5).
    In addition to addressing the needs of state and local gov-
ernments, the FLC members have increased their efforts toward
transfers to industry--both through encouraging secondary spinoffs
of technology as well as aiding in the commercialization of R&D
results.

SOME EXAMPLES OF FLC COMMERCIALIZATION ACTIVITIES

Team Unit Commercialization

    Approximately three years ago, one FLC member, the Naval
Ocean Systems Center (NOSC), was asked by the Naval Electronics
Command to design and build a microphone and earphone device which
could be used with a new radio being built for the Marine Corps.
The tube earphone and microphone (TEAM) was designed to allow the
radio operator to have hands-free communications by attaching a
small lapel-worn microphone and tube earphone device to the
operator's jacket.
    The NOSC Technology Transfer Office was made aware of this
device by the Navy inventor. Believing there to be some potential
commercial application in the law enforcement or construction
industry for such a device, the Technology Transfer Office began
to search for commercial companies which would be interested in
producing this product. Working with the California Innovation
Group (CIG), a contact was made with a small firm, Ramp Industries, Binghamton, New York. Technical information, pictures,
wiring diagrams, and parts lists were sent to Ramp Industries by
the CIG. (The CIG, now named the Southwest Innovation Group (SIG),
is the first geographically-based network of Science and Technology agents who work closely with local governments to help
apply innovative techniques to public management. Innovation
groups now operate in ten regions of the country. CIG was originally set up as a joint NSF-NASA project with four California
cities and "back-up" assistance from aerospace firms in these
cities.)
    NOSC was contacted for help in acquiring a prototype unit for
evaluation. This resulted in issuance by the Office of Naval
Research Patent Office of an exclusive license to Ramp Industries

for the manufacture of the TEAM unit for commercial sale. This example of technology transfer highlights the cooperative role of the Federal laboratories with small business by identifying patentable items developed within a Federal agency which have commercial applications.

Lightweight Body Armor

In 1973 the Army was asked by the Law Enforcement Assistance Administration (LEAA) to determine the feasibility of developing a garment that would protect important public officials. The product desired was to stop bullets fired by most handguns and be resistive to knife attacks. Additionally, it should be inconspicuous when worn and yet comfortable enough to be worn for a full eight hours. Technically this was a quantum leap beyond the armor available at that time and required extensive knowledge of (1) ballistic materials; (2) user needs in terms of wear, maintenance and comfort; (3) testing methodologies; and (4) the specialized equipment necessary for such R&D effort.

The program to develop lightweight body armor involved a novel collaboration of various agencies. The Army's Edgewood Arsenal and Natick Research and Development Center were responsible for medical testing, garment development, and material evaluation. The National Bureau of Standards of the Department of Commerce drafted test standards. The Aerospace and MITRE Corporations provided operation requirements and conducted field evaluations. Private industries were contacted for information on materials and fabrics; and the Federal Bureau of Investigation, International Association of Chiefs of Police and others provided user guidance. After the material and design were developed, information was disseminated to body armor manufacturers and law enforcement officials.

In the last year, the lightweight body armor has saved the lives of over 200 law enforcement officers across the country. Several states have passed legislation requiring all law enforcement officers to be provided with lightweight body armor. Forty companies now produce the armor. Nineteen new companies were created as a result of this technology. The research programs continue at the Federal level to develop new materials to defeat higher levels of handgun threats. Ultimately, several hundred thousand officers working in high crime areas will be protected from 90% of the handgun threat.

Spinoff Transfer and Commercialization of Laser Technology

In December 1975, the DOE's Lawrence Livermore Laboratory conducted a two-day symposium to launch a formal effort in laser technology commercialization. The purpose was to consolidate information and to transfer practical technology to industry from the Lab's laser fusion program. The technologies included optical components, greatly improved optical materials and processing techniques, and major advances in several supporting technologies (e.g., precision machining, fast-transient diagnostic systems, and large high-energy pulsed power systems). Several firms became proficient as suppliers but none had enough information or experiences to build complete high-power systems of the kind needed by Lawrence Livermore Labs or others engaged in laser fusion research. It was felt to be desirable to eliminate this gap by transferring the necessary technology to industry in order to foster a broader and stronger industrial base for laser technology of the future.

To fill this gap, Lawrence Livermore Laboratory prepared a special set of technical papers describing its current solid state technology; acquired legal and patent clearance for the public symposium; and provided standard agreement forms for use by companies seeking further information and assistance through continuing consulting arrangements with Lawrence Livermore Laboratory and its employees.

As a direct result of the symposium, previous experience with Lawrence Livermore Laboratory as a vendor, and subsequent exchanges of information, at least one of the attending companies successfully bid on delivery of a high-power laser amplifier system to a large research institution. Two or more of the other attending companies are expected to receive commercial subcontracts and to receive prime contracts for commercial systems. Nearly all of the companies attending the symposiums continue to use the person-to-person communication links opened up by the symposium. Thus, there has been a commercial innovation for laser technology and the indications are that the technology is continuing to diffuse in the marketplace from its own momentum (6).

Local Technology Action Centers

In another approach to serving industry through technology transfer, the FLC's Far West Region has begun a demonstration project to establish an active brokerage service within the industrial community. The city of Santa Clara, California (in the heart of "silicon valley") was selected as the first trial and demonstration site. A Task Force composed of representatives of the Santa Clara Chamber of Commerce, the local business community, the FLC, the Southwest Innovation Group and the City of Santa Clara was formed to develop a plan of action and to develop an operational structure.

The basic structure that emerged called for a Chamber representative to provide interaction between a representative of a business or industry who has need for technology and a representative of a Federal Laboratory. The Chamber would insure the backup between the supplier and the user. The project would not be passive but an active and useful program to fully utilize technical information and expertise.

The proposed technology service was featured in the Chamber's Industrial Newsletter and a special edition devoted entirely to its potential benefits and advantages was distributed. The project was also the main topic of the Chamber's Industrial Seminar held on September 13, 1978.

The project has now entered its six-month demonstration period with the city of Santa Clara's Science Advisor, Mr. Warren Deutsch, serving as the temporary point of contact. Personal and direct contact will also be made with prospective business clients. At the end of the trial period, the result will be evaluated and the permanent operation of the program considered.

The San Diego area has been tentatively identified as a second and larger test site with possible introduction of a similar project at the start of calendar year 1979. Preliminary discussions have been held with both private and public sector representatives to inform them of the nature of the effort and to keep them abreast of progress of the Santa Clara experiment.

SUMMARY AND CONCLUSIONS

The Federal Laboratory Consortium for Technology Transfer is an active, growing organization of R&D laboratories and centers whose members are committed to efforts to speed the flow of R&D results from the Laboratory to the marketplace. The FLC members seem to function most effectively as brokers, serving as face-to-face contact points between the Technology resources and the users.

Some successful transfers have taken place, and new approaches to technology transfer and commercialization are being investigated. Our experiences in this area have led to a number of conclusions about the necessary steps to ensure successful transfer of Federal Technologies to the private sector (7,8,9).

The most important conclusions are the following:

1. The technology transfer activity must be a full-time, fully funded and directed effort on the part of the technology source.
2. Without active, informed and enthusiastic technology receptors, transfer efforts will fail.
3. Technology transfer agents, in the field or in the office, must have access to a broadly based body of technical information and experience, such as the Federal Laboratory Consortium for Technology Transfer.
4. These agents must have the freedom and the motivation to aggressively seek opportunities and to respond satisfactorily and in a timely manner to all requests for assistance.
5. Person-to-person contacts, over a long period of time, between sources and receptors are essential.
6. Merely providing information in the form of reports is usually not sufficient to effect transfers. Often, additional development work (tailoring a solution to a

problem) and/or training the receptor in the use of a
technical fix is required.

7. The transfer of a technology will be completed when the
technology becomes generally accepted practice, or when
the technology is readily available in the marketplace.

8. The transfer of Federal Technologies is an integrating
process, involving considerable effort on the part of
the receptor as well as the source and sometimes involving assistance from other sources, receptors, or
technology "brokers."

9. Participation of industry early in the transfer process
is essential.

LITERATURE CITED

1. Investigative Report on "Utilization of Federal Laboratories,"
Hearing, Subcommittee of the Committee on Appropriations,
House of Representatives, 95th Congress Second Sesssion,
February 1978.

2. See, for example, National Aeronautics and Space Administration,"Spinoff 1978, An Annual Report," (U.S. Government
Printing Office Washington, D.C.) January 1978.

3. For information on the Navy's Technology Transfer Program,
write: Headquarters Naval Material Command (Code 03T2),
Arlington, VA 22217.

4. For information regarding the Technology Utilization Bulliten,
contact: Mr. D.W. Jared, Oak Rodge National Laboratory,
Union Carbide Corporation- Nuclear Division, P.O. Box X,
Oak R idge, TN 37830, Telephone: (605) 483-8611, Ext. 3-0121.

5. "Technology and the Cities," National Science Foundation,
D.W. Gottlieb, editor, U.S. Government Printing Office
Washington, D.C., Stock Number 038-000-00213-9.

6. Dorn, D.W., "Technology Transfer at Department of Energy
Laboratories-Selected Case Studies from the Lawrence
Livermore Laboratory," UCRL-80502, February 16, 1978.

7. Maninger, R.C., "Some Commercial Innovations from Technology
Transfers of Federal Research and Development," UCRL-78312,
July 16, 1976.

8. McMillan, J.R., Wheeler, D.J., Koons, M.E., and Read, A.M., "Factors Influencing the Transfer of Government Technology to the Private Sector," Union Carbide Corporation-Nuclear Division report UCC-ND-291, October 1974.

9. Miller, C.F., "Some Approaches to Transferring Federal Technologies to State and Local Governments: The Lawrence Livermore Laboratory Experience," UCRL-79558, June 2, 1978.

RECEIVED March 14, 1979.

ANALYSIS

Commercialization of R&D Results

ROBERT J. CREAGAN[1]

Westinghouse Electric Corp., Gateway Center, Stanwix St., Pittsburgh, PA 15222

More than $44 billion will be spent in the U. S. for research and development during 1978. The source of funds and implementers of R&D are indicated in figure 1 which shows that industry funds 43% of the R&D and performs 68%. Unfortunately, a high percentage of R&D results have a limited circulation, are carefully filed and their potential contribution to society is lost, or delayed to be reinvented in a later R&D program. The huge investment and potential benefits involved provide tremendous economic motivation for proper selection of R&D objectives, plus evaluation and commercialization of appropriate R&D results.

The term commercialization is used to designate the transition from R&D results to a product or service sold and used in economically significant quantities.

From a charter and funding standpoint no organization has more motive for commercializing R&D than the Department of Energy. DOE's commercialization committee has analyzed various R&D technologies for five months to determine which were ready for commercialization. The results are summarized in figure 2.

To quote "Dale Myers" October 4, 1978 memo to "Jim Schlesinger" in part: "...In general where R&D for a technology is 'complete,' responsibility for 'marketing' activity is transferred to the Assistant Secretary responsible for commercialization...A Resource Manager is appointed to provide a DOE-wide point of focus for the integration of all activities required to achieve the earliest date for commercialization. ...Where large technology demonstrations are still needed, or the technology is not ready for commercialization because of cost or other barriers, we have not transferred the technology."

A second series of task forces will examine commercialization potential of six technologies identified by the Domestic Policy Review Group and listed in figure 3.

Projections by DOE in March of 1978 for commercial products

[1]Analysis prepared October 1978

0-8412-0507-8/79/47-105-097$05.00/0

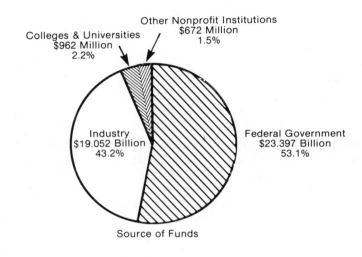

Source of Funds

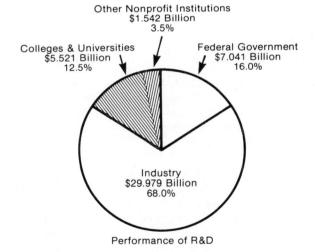

Performance of R&D

Figure 1.   Total U.S. R&D in 1978 $44,000,000

Electric Market

*Hydrothermal/Geothermal Generation
*Low Head Hydro Generation
*Small Wind Generation
  **Combined Cycle with Integrated Gasifier for Utility
    Application
  **Fuel Cell Power Plants
  **Large Wind Generation
  **Atmospheric Fluidized Bed Combustion for Utility
    Application
  **Photovoltaics--to be resolved November 15, 1978

Liquid Fuels

*Enhanced Oil Recovery
*Oil Shale, Surface and In-Situ Retorting
  **Coal Liquification

Gaseous Fuels

*Enhanced or Unconventional Gas Recovery
*Low-BTU Coal Gasification
*High-BTU Coal Gasification (first generation only)
  **High-BTU Coal Gasification (Advanced Technology)
  **Medium-BTU Coal Gasification

Direct End Use Applications

*Cogeneration
*Conservation Products Marketing (oil burner retrofits, high
 high efficiency motors, air fuel ratio)
*Electric and Hybrid Vehicles (first and second generation)
  **Electric and Hybrid Vehicle (third generation-hot batteries)
*Passive and Hot Water Solar Heating
*Urban Waste

 *Ready for commercialization, effective September 30, 1978, DOE
**Not ready

Figure 2.   R&D technology reviewed by DOE for commercialization

1. Wood Combustion (for both industrial and utility application)
2. Solar Industrial Process Heat (including use of solar energy
   to generate steam for enhanced oil recovery projects)
3. Non-Battery Storage Facilities for Utilities (including
   underground compressed air and underground hydro)
4. Annual Cycle Energy Systems (a planning system based on
   energy that could be derived from natural sources, such as
   wind or water, that change from season to season)
5. Lighting Efficiency
6. Thermally Activated Heat Pumps

Figure 3.   R&D technologies to be reviewed by DOE for commercialization

available in 1985 and 2000 from commercialization of R&D results are given below in terms of thousands of barrels per day of oil equivalent.

DOE ENERGY PRODUCTION ESTIMATES
(in thousands of barrels/day of oil equivalent)

	1985	2000
Enhanced Oil Recovery	600-1,000	2,200-4,500
Enhanced Gas Recovery	500-1,600	1,400-3,700
Heat Engines & Heat Recovery*	500-1,000	1,500-3,000
Low-Btu Gasification	50-250	400-950
High-Btu Gasification	50-200	900-1,800
Direct Combustion	40-150	400-750
Oil Shale	40-130	800-2,000
Liquefaction	20-80	800-1,200
Fuel Cells	10-40	30-100
In Situ Coal Gasification	0-15	80-150
Advanced Power Systems	0	0-50
MHD	0	0-20
TOTAL	1,810-4,465	8,510-18,220

*through conservation efficiency

At $20 per barrel the production estimate total is worth up to $33 billion in 1985 and $133 billion in the year 2000 compared with 1978 total R&D expenditures of $44 billion which includes R&D for many other items. Hence if the projections are correct, the invested R&D funds will be exceeded by commercial sales during each year of production after 1985.

An historic example is the $2 billion spent essentially for R&D on the Manhattan Project from 1940 to 1945 and the resulting worldwide nuclear power industry with 408,285 MWe operable, under construction or on order as of June 30, 1978. These nuclear power plants, at $500/kwe represent a capital investment of over $200 billion, plus expenditures for fuel, at 20 mills/kwh, of $50 billion per year.

These examples illustrate huge potential benefit/cost ratios which can be obtained from commercialization of R&D. What may not be appreciated is that the costs of commercialization are usually much greater than the costs of R&D and hence commercialization requires more decisions and efforts than R&D.

For perspective the curve shown in figure 4 can be used as a typical financial forecast of net cash flow after taxes resulting from expenditures and income associated with performing R&D, and eventually making a profit by commercializing the R&D results. The cash flow amplitude and time dimensions of the curve change but the shape is similar whether it is a large program such as commercializing nuclear power or a smaller project such as a superconducting generator or a high voltage bushing.

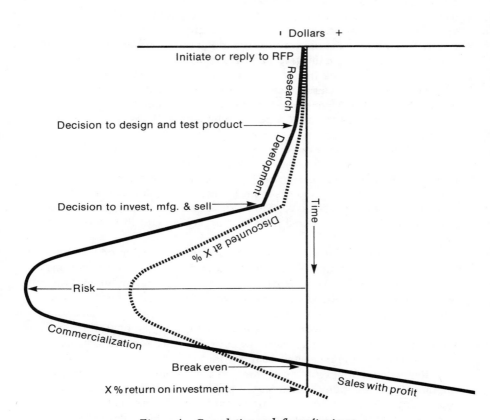

Figure 4.   Cumulative cash flow after taxes

Cash flow typically is comparatively small during the R&D period but increases as development and manufacturing hardware and investment is involved.

Critical milestone decisions are: (1) Do company objectives motivate the R&D? (2) Are research results favorable enough to justify continuing into a development phase during which prototype project will be manufactured and proven in operation? (3) Do market evaluations show that enough product can be sold at a profitable price to warrant investment for manufacturing facilities, marketing expenses and other financial commitments. A decision to commercialize the R&D will lead to a maximum negative cash position indicated in figure 4 as "risk" before cumulative profits are available from increased production to pay back investment and potentially make a profit.

Typically the net cash flow curve is discounted at a target return-on-investment (ROI) percentage and when the cumulative discounted cash curve crosses the zero line the business will have an ROI equal to objective. (See figures 4 and 9.)

Evaluating a predicted discounted net cash flow curve will determine to a large extent whether an R&D result is worth commercialization. Uncertainties exist and have been incorporated in some analyses by increasing the discount percentage by an increment proportional to uncertainty. However, a prudent management must accept uncertainty risks and periodically reevaluate the decision to proceed.

A cash flow curve for R&D plus commercialization of a transmission voltage bushing is shown in figure 5. EPRI has funded this project since 1975. Total R&D costs, including corporate cost sharing, will be about $700,000 when the project is complete this year. In comparison, forecast commercialization costs involve a temporary negative cash flow of over $5 million. Details of the cash flow are tabulated in figure 6. This table is provided only for illustration because realistic market and cost estimates are not yet available.

Non-recurring strategic commitments include funds for planning, engineering, designing, manufacturing, land, building, machinery and marketing. The negative cash flow starts turning around in 1982 because of depreciation tax credit adjustments which can be netted against profits made elsewhere in the corporation even before any income is received from sales. Sales billed from 1983 on, with an assumed IBT of 24% on sales, provide potential for investment payback by 1986. Return on investment for the project for the 10 years from 1978 through 1987 is 12.5% if R&D costs are not included, because they were paid by EPRI. If R&D costs of $700,000 are included, ROI is reduced to 10.4%. Such a low ROI forecast provides little incentive for taking much risk in commercialization but the 12.5% ROI exclusive

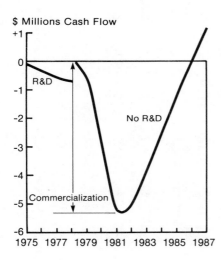

Figure 5. Commercialization of R&D cash flow

	1975	1976	1977	1978	1979	1980	1981	1982	1983	1984	1985	1986	1987
R&D	100	250	250	100									
Engineering				100	150	100	50	50					
Manufacturing					50	50							
Plans and Designs						300	300						
Marketing							50	50					
Land					500								
Building						500	500						
Machinery & Equipment						2000	2000						
Sales Billed									8400	8400	8400	8400	8400
IBT (Before Dep.)									1982	1982	1982	1982	1982
IRS									952	952	952	952	952
Tax Adjustments				48	96	521	686	395	298	259	221	183	145
Net Cash Flow (No R&D)				-52	-604	-2429	-2214	295	1328	1289	1251	1213	1175
Cum Cash Flow (No R&D)				-52	-656	-3085	-5299	-5004	-3676	-2387	-1136	+77	+1252

Terminal Value 3120

ROI (No R&D) 12.5%
ROI (With R&D) 10.4%

Figure 6.   Commercialization of R&D (thousands of dollars)

of R&D might be marginally acceptable if a product line is rounded out or cost reduced with complete assurance of success.   The difference in ROI illustrates the benefit of outside funding for R&D costs which are weighted most heavily in a discounted cash flow because they are up-front costs.   EPRI or DOE funds thus allow a company to investigate more possibilities or carry out more thorough analysis before a decision is made to commit the larger funds for commercialization.

From an EPRI or DOE investment viewpoint, R&D expenditures can be justified in many cases where a manufacturer would not invest because the ROI calculated by EPRI or DOE is greater.   EPRI's or DOE's more favorable ROI may result from two factors i. e. no payment for commercialization costs, and/or the fact that all the benefits and hence income to utilities or society accrue to EPRI or DOE while only a fractional market capture accrue to the competitive manufacturer. Thus with a better ROI, EPRI or DOE has more motive to pay for R&D than the manufacturer even if cost of money were the same to both.   In addition money may be less costly to EPRI or DOE than it is to the manufacturer hence a lower discount can be used for cash flow and longer time between investment and payoff can be acceptable.   Such financial factors can explain to some extent logic which makes govern- ment investment in fusion R&D tenable, while a manufacturer could not endure the decades of negative cash flow before a profit is even possible.

In considering what R&D results should be commercialized, it is not enough for a manufacturer to consider income after taxes, because high investment requirements in a capital intensive business may require negative cash flow for an extended period even with a positive IAT, especially if rapid growth is involved.   For example using financial ratios of 4% after tax profits on sales, and investment of 37% of sales, a growth rate greater than 15% in sales would result in a negative cash flow.

The effect of investment requirements for commercializing a superconducting generator is illustrated in figure 7 where the curve of cumulative IAT which is a measure of ability to generate cash, is contrasted with investment curves based on different investment assumptions as follows:

Case 1.   A completely new business is assumed with financial ratios of working capital plus inventory at 29% of sales, and plant plus equipment at cost as 42% of sales.

Case 2.   Case 1 above is modified to reduce capital for plant plus equipment to 21% of sales assuming that the stator, which is about half the cost of the generator, would be manufactured in existing facilities while new facilities would be required only for the superconducting rotor, hence only half the investment for new plant and equipment.

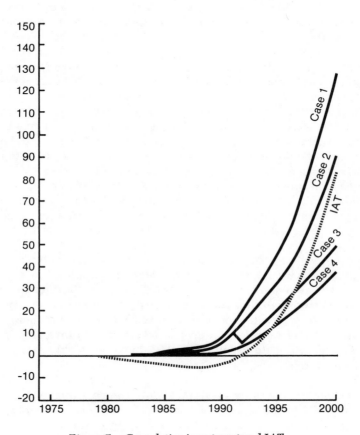

Figure 7.   Cumulative investment and IAT

Case 3. Case 2 above is modified to reduce working capital plus inventory fund requirements by assuming that such funds partially would be made available as conventional generators are phased out and their associated investment funds are released.

Case 4. Case 2 above is modified so that working capital plus inventory fund requirements are assumed to be zero based on progress payments from customers.

Cumulative cash flows which result from the above four investment cases are given in figure 8 respectively as follows:

Case 1. The negative cash flow cumulatively increases to $29 million in 1998 and starts to turn around because of positive IAT in 1999 and 2000 but is still a poor business prospect because of large investment requirements compared to IAT.

Case 2. Based on lower investment requirements because of use of existing equipment and plant for stator, the cumulative negative cash flow increases to $15 million in 1993, turns around and is cumulatively positive $5 million in the year 2000.

Case 3. Assuming use of capital funds released by phase-out of conventional generators which is what actually would happen, the cumulative negative cash flow increases to $11 million in 1991, turns around and is cumulatively $46 million positive by 2000, with a 10% ROI by 1998 and a 15% ROI by 2000 without terminal value in either case.

Case 4. Customer progress payments to decrease working capital plus inventory investment to zero, limit cumulative negative cash flow to $6 million in 1988 and permits cumulative positive cash flow of $58 million by the year 2000 for 10% ROI in 1996 and a ROI of 15% by the year 1998 without terminal value in either case. A turn around in cash flow occurs in 1988, three years after the first unit is on line, and a break even on cumulative cash flow occurs in 1994. This example makes a strong case for progress payments and developments in the manufacturer's existing facilities in order to minimize investment requirements.

For commercialization it is not enough that an R&D program be a good thing for the manufacturer, it must also provide value to a customer. The curves in figure 9 show the cash flow for utilities in terms of supporting R&D ($24 million) for a superconducting generator and the benefits which they receive. Benefits are based on a capitalized evaluated worth of $25/kwe for a superconducting generator because of its greater efficiency (.8%) and the favorable cost reduction for total power plant capital, fuel cost, plus operation and maintenance. Utility cash flow benefits are based on reduced revenue requirements at 10% based on reduction of utility capital requirements because of the $25 per KWe capitalized benefit of the superconducting generator. Return on investment for the utilities is 15% by 1992 with an expenditure of less

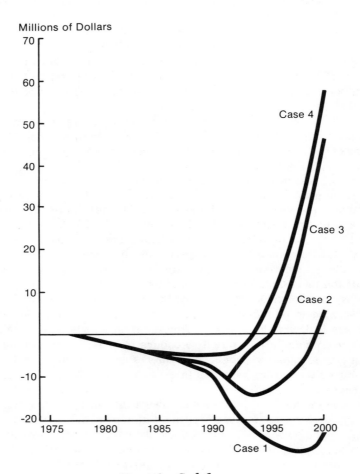

Figure 8.   Cash flow

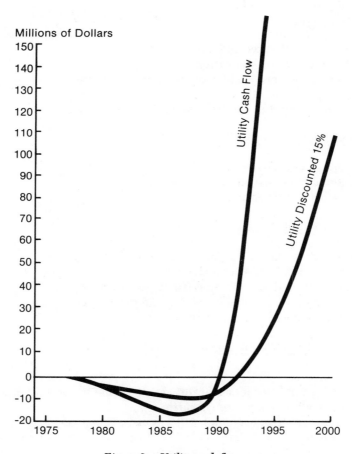

Figure 9.   Utility cash flow

R&D Technology Reviewed by DOE for Commercialization

Electric Market

*Hydrothermal/Geothermal Generation
*Low Head Hydro Generation
*Small Wind Generation
　　**Combined Cycle with Integrated Gasifier for Utility
　　　Application
　　**Fuel Cell Power Plants
　　**Large Wind Generation
　　**Atmospheric Fluidized Bed Combustion for Utility
　　　Application
　　**Photovoltaics--to be resolved November 15, 1978

Liquid Fuels

*Enhanced Oil Recovery
*Oil Shale, Surface and In-Situ Retorting
　　**Coal Liquification

Gaseous Fuels

*Enhanced or Unconventional Gas Recovery
*Low-BTU Coal Gasification
*High-BTU Coal Gasification (first generation only)
　　**High-BTU Coal Gasification (Advanced Technology)
　　**Medium-BTU Coal Gasification

Direct End Use Applications

*Cogeneration
*Conservation Products Marketing (oil burner retrofits, high
　efficiency motors, air fuel ratio)
*Electric and Hybrid Vehicles (first and second generation)
　　**Electric and Hybrid Vehicles (third generation-hot batteries)
*Passive and Hot Water Solar Heating
*Urban Waste

*Ready for commercialization, effective September 30, 1978, DOE.
**Not ready.

R&D Technologies to be Reviewed by DOE for Commercialization

1. Wood Combustion (for both industrial and utility application)
2. Solar Industrial Process Heat (including use of solar energy to generate steam for enhanced oil recovery projects)
3. Non-Battery Storage Facilities for Utilities (including underground compressed air and underground hydro)
4. Annual Cycle Energy Systems (a planning system based on energy that could be derived from natural sources, such as wind or water, that change from season to season)
5. Lighting Efficiency
6. Thermally Activated Heat Pumps

than $25 million for R&D and in return benefits worth more than a billion before the year 2000. This is an example of an R&D project that provides over 15% return on investment to both manufacturer and utility customer, is beneficial for all parties involved, and therefore undoubtedly will be commercialized.

Comments and Conclusions

1. Commercialization of results is one of the most important objectives in R&D management.
2. Selection of appropriate R&D projects for commercialization is not subject to exact analysis because probability for success is not established.
3. Evaluation of R&D results for possible commercialization should include consideration of IAT, cash flow, incremental investment requirements and fit to present company facilities, markets, technology, and management.
4. The R&D results which offer greatest opportunity for commercialization in terms of ROI are in the area of existing business because of management capability, customer contacts, existing investment in pertinent facilities and equipment and experienced staff.
5. Working backwards from the market place in terms of what your customers need or want are useful considerations in selecting appropriate R&D for proposals or for commercialization.
6. For business programs with capital intensive investment requirements, customer progress payments are highly desirable.
7. Decisions as to whether an R&D project can be commercialized successfully should be explored during the R&D phase so pertinent technical and economic uncertainties can be studied and commercial objectives can be established.

8. Long term or high risk R&D programs with vague payback times should be funded by DOE or EPRI if, assuming success, they provide significant benefit to society or utilities.

9. Excellence of implementation is often more significant to economic success than marginal theoretical benefits in commercialization of R&D results.

10. Commercialization is often more costly than the R&D involved.

RECEIVED March 14, 1979.

Commercialization and the Assessment of Federal R&D

GEORGE TOLLEY and STUART TOWNSEND

University of Chicago, Department of Economics, Chicago, IL 60637

Research has provided us with many new processes and ideas
that can be utilized in new technologies for the production and
conservation of energy. Many of these potential technologies are
not profitable at present. However, rising fuel prices, improved
efficiencies and reduced production costs will increase profitability until many of the more exotic technologies will, at some
time, become economically feasible. Research and development
projects can accelerate this process. Research and development
projects can be analyzed as investment activities where the
return to the R & D is an earlier stream of benefits from a new
technology.

Serious assessment of the potential of any given specific
innovative activity is possible, but too seldom attempted. The
effects of a potential innovation on quality of life, the reductions brought about in costs of producing goods and services, the
extent of the markets and the rapidity of adoption, as well as
the cost of the R & D activity can all be estimated, at least to
an order of magnitude. This type of assessment can be used as a
basis for encouraging or discouraging particular types of activity. This paper develops a methodology for determining which of
the many R & D opportunities available offer the greatest potential returns to society.

The first section introduces and elaborates on the four
stages of the proposed methodology. The second section introduces interdependencies among projects relevant to the selection
of portfolios of projects and illustrates methods of selecting
portfolios. The third section develops the mathematical basis
for computerizing the methodology and applies it to the case of
solar heat for residences. The fourth section presents conclusions.

Description of the Proposed Evaluation Method

The purpose of the remainder of this paper is to develop a
method for the systematic comparison of energy storage R & D

0-8412-0507-8/79/47-105-113$05.00/0

projects which are competing for Federal funding. The objective
of the government in deciding which R & D projects to fund
should be to obtain the most return from public investment in
energy-conserving technologies.

The complete R & D project evaluation and selection procedure should deal with:

1. The likelihood of commercial success, taking into
   account:
   a. the potential application of the product
   b. the size of the potential markets for the product
   c. the existence of competing products and elastici-
      ties of substitution between products
   d. the cost of manufacturing and marketing the
      product
   e. the return on the manufacturer's investment and
      to consumers
   f. the timing of the manufacturer's introduction,
      rates of penetration, and market saturation of the
      product
2. An analysis of the technical soundness of the project
   including:
   a. the probability of technical success
   b. the probability of achieving product unit-cost/
      performance objectives
   c. project development costs and schedules
3. The need for and effects of the Federal R & D support,
   considering:
   a. the existence of similar R & D projects in the
      private sector
   b. the reasons for the lack of sufficient private
      sector R & D
   c. the acceleration of commercial deployment due to
      Federal expenditures
4. The potential energy resource savings to the nation,
   including:
   a. fuel oil and natural gas savings
   b. other resource (capital, labor, and material)
      savings

The methodology developed to integrate the consideration of
these issues into a consistent measure of project worth is a
nested cost-benefit analysis. Figure 1 illustrates the four
stages of the methodology.

The first stage of the analysis determines the private cost
savings produced by a single unit of the new innovation installed in a particular time period and location. The private
cost savings are measured by the difference between the present
values of the total cost of the old and new technologies. The
second stage of the methodology uses the projected private cost
savings per unit to estimate the rate and ultimate extent of
market penetration as a function of the present value of the

savings and other market factors. If it were possible to define
homogeneous markets, and to calculate the private cost savings
for each, then market penetration would be a dichotomous decision.
If the innovation would result in lower costs, it is installed;
if higher costs would result, it is not installed. Since all
firms, or households, in the homogeneous market are alike, either
all or none adopt. In practice it is not possible to define completely homogeneous markets and it is necessary to estimate the
rate and ultimate extent of market penetration. Factors, in
addition to average cost savings, which affect market penetration
for the typical firm include the scale of the firm relative to
the innovation, the vintage of the firm's capital equipment, the
age and education of its management, and the expected growth of
the industry.

Figure 2 illustrates one functional form used to estimate
the rate and ultimate extent of market penetration, the logistic.
The logistic may be expressed as

$$Y_t = \frac{a}{1+b\ e^{-rt}}$$

where: a is the ceiling level as a percent of the total
        b is a constant related to the initial level
        r is the rate of adoption
        Y_t is the percent of the total who have adopted by year t

        and t represents time.
When t is equal to zero the percent who have adopted is equal to
a/1+b and when t is equal to infinity the percent who have
adopted is equal to a.

The projection of market penetration is the most tenuous
link in the process of determining the expected benefits of an
R & D project but some estimate must be made. The private sector makes these kinds of projections every day and must live with
the outcomes. If the government is to carry out an effective
R & D program directed toward the ultimate commercialization of
innovations, it must also develop the capability of making reasonable projections of future markets.

The third stage of the methodology computes the present
value of the social cost savings per unit by replacing the private prices, used in stage one, with prices which have been
adjusted to account for market failure and externalities. The
expected present value of the social cost savings per unit for
each time period is multiplied by the projected number of adopt-
ers in that period. This series is then discounted and summed to
yield the total expected present value of the social cost savings
generated by the new innovation.

This stage of the methodology is very important since, for a
variety of reasons, the prices charged to individual consumers
often do not reflect the true cost to society of various inputs.

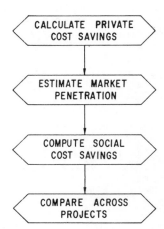

Figure 1.   Evaluation procedure

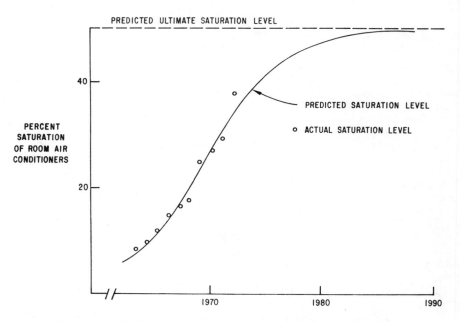

Figure 2.   Market penetration for room air conditioning in Wisconsin

This may be due to regulatory distortions in the market or to
the presence of externalities, i.e., costs which are not taken
into account by the consumer because they are not reflected in
the price he must pay. Most often these externalities are such
things as pollution, which impose costs on those who are exposed
to them even though these people may not be receiving the benefit
of the product associated with the pollution.

In the market for energy, the main causes of divergence
between private and social prices are the regulation of natural
gas prices, the existence of a social cost associated with U.S.
dependence on imports of foreign oil not reflected in private
prices, and the existence of environmental externalities such as
pollution or risk of a major accident such as nuclear leakage.

The capital market in the United States is distorted by the
corporate income tax which increases the before tax return to
corporate capital and reduces the before tax return to capital in
the remainder of the economy from the level which would prevail
in the absence of the tax. While the private rates of return on
capital are used to estimate the expected present value of private
cost saving, the social opportunity cost of capital is used to
evaluate the expected present value of the social cost savings.

The fourth stage of the methodology compares the expected
present value of the total social cost savings to the present
value of the R & D expenditures required to realize the innovation. Ratios, such as the net benefit to government R & D cost
ratio, can be used to compare the worth of various alternative
projects.

This procedure, however, is correct only when there are no
interactions among projects. The third section of this paper
discusses selection procedures which should be used when there
are interdependencies among projects.

Portfolio Analysis

The evaluation process previously described treats a single
R & D project and a single existing alternative. In reality
there are many R & D projects each competing for several end
uses. The correct unit of analysis is not the individual project, but rather the portfolio of projects controlled by the program manager. The interactions among projects during the R & D
process and in the market place must be taken into account. A
very simple example will illustrate the principles involved.

Suppose, that two competing technologies exist, should both
be funded? Assume:
1. the two projects considered separately have benefits
    greater than costs;
2. if one is successful, the other will have no benefits.

For instance, the two projects could be alternative life support systems for space men or competing energy storage devices

for use in 1980. The question is should one project, or the other, or both be funded.

Let the expected benefits from supporting each technology separately be given by

$$E_1 = VP_1 - C_1$$

$$E_2 = VP_2 - C_2$$

where V is the energy cost savings (same for both technologies), P is the likelihood of the technology coming on line in the target year, C is the cost of developing the technology.

The expected benefits if both technologies are supported is given by:

$$E_3 = E_1 + E_2 = V(P_1 + P_2 - P_1 P_2) - C_1 - C_2$$

(The term $P_1 + P_2 - P_1 P_2$ results from the rules for summing two probabilities which states that the probability of a sum is the sum of the probabilities minus the covariance.)
E_3 is greater than E_1 if

$$V(P_1 + P_2 - P_1 P_2) - C_1 - C_2 > VP_1 - C_1$$

$$VF_2 - VP_1 P_2 - C_2 > 0$$

$$VP_2 - C_2 > VP_1 P_2$$

(1) $E_2/V > P_1 P_2$

Similarly E_3 is greater than E_2 if

$$V(P_1 + P_2) - C_1 - C_2 > VP_2 - C_2$$

or

(2) $E_1/V > P_1 P_2$

Thus if E_1/V and E_2/V are greater than $P_1 P_2$ the net benefits for proceding concurrently exceed the benefits from proceding on either project individually. A brief numerical example will illustrate the principle. Let V = $100 million, C_1 = $1 million, and C_2 = $5 million. Suppose the risk of not meeting the target year is the same for both technologies and equal to .2. Then $P_1 = P_2 = .8$. For the first condition (1) we have

$$\frac{E_2}{V} > P_1 P_2$$

$$\frac{100 \cdot .8 - 5}{100} > (.8)(.8)$$

$$.75 > .64$$

For the second condition (2), we obtain

$$\frac{E_1}{V} > P_1 P_2 \quad , \quad .79 > .64$$

Thus both conditions are fulfilled and both technologies should be supported.

A detailed budget of the R & D project is required to facilitate the evaluation of separable components of the proposal. This will allow partial funding of projects where either budget constraints or overlapping projects indicate that full funding is not possible. The general rule for analyzing separable components of a project is that each component should be funded if, when evaluated at the appropriate discount rate, it generates positive expected net present values. Where differential funding levels of a component are possible the incremental return from the last increase in funding should just equal the opportunity cost of those funds. Following these rules will maximize the net present value of the project.

The principles illustrated by this example may be generalized for more complex situations.

Application

This section develops the algebraic model of the methodology which was introduced in the first section, and describes a computer program which performs the evaluations, and presents the results of a sample evaluation of solar heating for single family residences.

The algebraic expressions for the private costs of the new and old technologies are shown in Figure 3. The term $B(t-t_1+1)$ is the discount factor, based on the private opportunity cost of capital. The term $\bar{p}$ is a collection of vectors, one for each time period and location couple (i, t), that contains the prices, for the specific time and location, of all inputs (capital, fuel and labor). The terms $\bar{q}_n$ and $\bar{q}_u$ are collections of vectors that

PRIVATE COST WITH OLD TECHNOLOGY

$$V_0(i,t) = \sum_{t=t_1}^{t_1+T} B(t-t_1+1) \; \overline{p}(i,t) \; \overline{q}_0(i,t)$$

PRIVATE COST WITH NEW TECHNOLOGY

$$V_N(i,t) = \sum_{t-t_1}^{t_1+T} B(t-t_1+1) \; \overline{p}(i,t) \; \overline{q}_N(i,t)$$

NUMBER OF UNITS OF THE NEW TECHNOLOGY

$$M(i,t_1) = F\,[V_0(i,t_1) - V_N(i,t_1),\; \overline{R}(i,t_1)]$$

Figure 3

contain the input requirements for the old and new technologies respectively. The $\bar{q}$ vectors are ordered to correspond to the $\bar{p}$ vector. $V_o(i, t_1)$ is the present value of the private cost associated with one unit of the old technology installed in location i and time period t_1. $V_n(i, t)$ is the present value of the private costs associated with one unit of the new technology installed in location i and time period t_1.

Market penetration, $M(i, t_1)$, is a function of the present value of the private cost savings and other factors. The example described below uses a logistic function for F and the other factor considered is the projected number of new housing starts in each location and time period.

Figure 4 presents the social costs of each technology and the formula for discounting and summing the social cost savings. The asterisk indicates that the prices and the discount factor reflect the social value of inputs and the social opportunity cost of capital. The total social benefits are obtained for 65 locations and for all years in the forty year planning period. The 65 locations are the 65 largest Standard Metropolitan Statistical Areas (SMSA's). A computer program has been developed to perform these calculations for several types of technologies. The primary computer language used was PL/1 with provisions for using FORTRAN in some of the cost calculations if desired.

Figure 5 shows the elements of the system. The system resides on an IBM 370/168 in a TSO environment. Control lists (CLISTS) are provided for the Editor and the Evaluator. A single data base contains all the data for prices, markets, and climate for the SMSA's. A data base Editor is provided for maintaining the data base and for producing reports of the information in the data base.

The Evaluator is stored in a library of object modules. When the Evaluator CLIST is executed the user specifies which technology he wishes to evaluate and TSO constructs the desired program and executes it. The program then prompts the user for a variety of information on costs. Once the calculations are completed the user specifies the type of reports he wishes and the program produces them. Figures 6, 7, and 8 outline these steps.

The evaluation methodology described in the previous sections has been implemented for solar heat for single family residences. It should be emphasized that this analysis is a prototype of a methodology for the evaluation of R & D projects and is not meant to be a thorough evaluation of the future of solar energy. The particular example chosen is the evaluation of a sample R & D program designed to reduce the energy storage costs of residential solar applications from $4 per square foot of collector to $2 per square foot between 1980 and 2000 and to $1 per square foot between 2005 and 2020 and to increase the

SOCIAL COST WITH OLD TECHNOLOGY

$$V_0^*(i,t_1) = \sum_{t=t_1}^{t_1+T} B^*(t-t_1+1)\bar{p}^*(i,t)\bar{q}_0(i,t)$$

SOCIAL COST WITH NEW TECHNOLOGY

$$V_N^*(i,t_1) = \sum_{t=t_1}^{t_1+T} B^*(t-t_1+1)\bar{p}^*(i,t)\bar{q}_N(i,t)$$

SOCIAL BENEFITS OF NEW TECHNOLOGY (ONE REGION, ONE YEAR)

$$\omega(i,t_1) = M(i,t_1)\, V_0^*(i,t_1) - V_N^*(i,t_1)$$

Figure 4

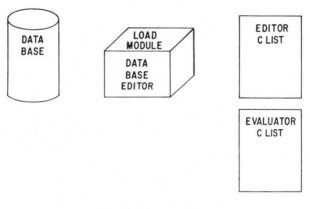

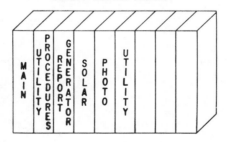

Figure 5.   The system

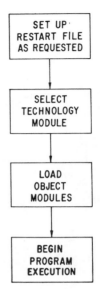

*Figure 6.   Preprocessor (TSO command
language)*

Figure 7.   Evaluation program (PL/1)

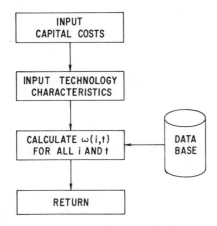

Figure 8. *Technology module (Fortran or PL/1)*

Table I

SOCIAL BENEFITS TO RESIDENTIAL SOLAR APPLICATIONS
WITH AND WITHOUT AN R & D PROGRAM IN ENERGY STORAGE

Year	Benefits with Storage R & D	Benefits without Storage R & D	Benefits to Storage R & D
1980	1.2	.2	1.0
1985	1.4	.2	1.2
1990	.8	.1	.7
1995	.4	.1	.3
2000	.8	.2	.6
2005	.6	.2	.4
2010	.4	.1	.3
2015	.3	.1	.2
Total:	5.9	1.2	4.7

lifetime from 10 to 20 years for all periods. The solar technology supplying energy is a water-cooled collector costing $8
per square foot between 1980 and 2000 and $4 per square foot
between 2005 and 2020. The lifetime of the solar collector is 20
years. Solar energy supplies 50% of the heating requirements of
the home and also supplies domestic hot water during the non-
heating season.

The private price projections used are those implied by the
National Energy Plan of 1977 as expressed in various DOE reports.
The social prices were derived from the private prices by cor-
recting for taxes and subsidies and by adding pollution and
foreign dependence costs where appropriate. The homeowner's real
opportunity cost of capital was 1% and the real social opportunity of capital was 8%. Table I shows the public benefits of the
solar technology with and without the storage R & D activity for
each time period. The present value of the total benefits of the
storage R & D activity under these circumstances is $4.7 billion.

This value must be considered in the context of the uncertainty of actually achieving the postulated cost reductions and
the potential erosion of the market by rival innovations.

Conclusions

The range surrounding these estimates must be stressed. The private and social cost savings from an improvement in technology depend on uncertain technical factors and future prices. These estimates are used to predict market penetration, which, again, has a random element. Social and private prices will tend to be correlated. Market penetration is largely a function of the private cost savings brought about by the improved technology. Errors in the estimation of private savings will thus lead to errors in both market penetration and social savings. It is important as part of any use of the method that a range of values be considered and the sensitivity of the results be determined.

In summary, the evaluation of Federal R & D programs for the commercialization of new technologies requires a four-step analysis. The first and second steps, cost evaluation and market analysis, mimic the analysis that a private firm would perform prior to embarking on a new venture. The third step, the calculation of social cost savings, is unique to the government's point of view. Large firms, such as conglomerates, and the government must also consider the fourth step, portfolio analysis, in making programatic decisions. The procedure proposed is no more accurate than the data available but does provide a logical method for incorporating diverse information into a consistent decision-making process.

RECEIVED March 14, 1979.

Assessing the Government Role in the Commercialization of Federally Funded R&D

ALDEN S. BEAN

National Science Foundation, Division of Policy Research & Analysis, Washington, DC 20550

J. DAVID ROESSNER

Solar Energy Research Institute, Policy Analysis Branch, 1536 Cole Blvd., Golden, CO 80401

Since the second world war we have observed a continuing expansion of the Federal government's efforts to influence the rate and direction of technological change in the civilian economy. In the immediate post-World War II period, nonmilitary basic research was supported for a variety of reasons including national prestige, investment in a store of knowledge from which future technology could draw, and variations of the "market imperfection" argument. The dramatic increase in Federal funding of R&D immediately following Sputnik also stimulated Congressional interest in maximizing the public's investment in R&D, particularly defense and space R&D; this interest was manifested in mandates to disseminate the results of government-sponsored R&D to the civilian economy. Then information dissemination and demonstration programs were initiated as elements of the numerous social service programs that grew and multiplied in the 1960's. Now, in the 70's, information dissemination and demonstration have become integral parts of civilian R&D programs focusing on hardware technology as well. The most recent expansion of Federal government activity is labeled "commercialization," where any mix of strategies (R&D, demonstrations, information dissemination, subsidies) may be employed to facilitate or stimulate technological change in the civil sector. (A major exception to this pattern is, of course, Federal support via cooperative arrangements with industry to develop the light water nuclear reactor. However, this program was directed toward a single technology rather than toward the broader goal of technological change.) With this latest expansion have come increased government involvement with, and dependence upon, industrial research and technological development activities if government objectives are to be achieved.

This paper focuses on research approaches and methods used to assess the effectiveness of government programs that are intended to achieve these objectives. The overriding purpose of this paper is to discuss the research methods that have been or might be used to best advantage in assessing policy issues associated with the commercialization of Federally funded R&D. This paper begins by

0-8412-0507-8/79/47-105-129$05.00/0

describing several policy issues that characterize the debates
surrounding decisions to support commercialization programs with
public funds.

Policy Issues in the Commercialization of
Federally Funded Developments

The policy issues inherent in decisions to spend public funds
on commercialization are perhaps most sharply drawn in the negotiations between government budgeting executives and technology
transfer program directors. During these negotiations, day-to-day
operating problems are often set aside and fundamental assumptions
justifying the programs are likely to be reviewed and questioned.
These include questions of whether a government sponsored technology transfer initiative is warranted at all; if so, how incentive mechanisms should be structured so that the benefits, costs
and risks associated with the effort are equitably distributed and
shared; and in determining the time frame or sequence of events
over which the government's involvement should last. The public
administration literature does not provide generalizable answers
to these questions. However, the literature does suggest some
general principles to guide government decisions about the investment of public funds in activities intended to enhance the long
range productivity and economic growth of the nation (1). Since
these general principles have been widely accepted as providing a
rationale for investments in basic and applied research, it seems
desirable that the rationale for public investment in technology
commercialization activities should also be consistent with them.
The general principles include the following assumptions (see (2)
for a similar discussion).

1. The benefits to society of a Federal investment
should exceed the costs imposed on society by that
investment.

2. The net benefits or net return from the investment is at least as great as the net benefits from
alternative investments the government could make
(opportunity costs).

3. There are inadequate incentives for industry
to undertake the proposed investment if government
does not.

4. There is an absence of institutional changes
that could be made to stimulate the needed investment.

These criteria are based on notions of efficiency with which
most people would agree. That is, government should invest
society's resources in areas which have the highest payoffs to
society. There are, however, cases where it is difficult to
project in a quantitative fashion a future benefit to society.
Moreover, there have been cases where the Federal Government has
made investments in the country's future where no payoffs were

expected in the foreseeable future, but where long range, unexpected benefits have accrued. While historically these have been
exemplified mainly in such investments as land purchases--the
Louisiana Purchase and the Alaska purchase--there have been
unexpected benefits from investments in technologies in the
nuclear, military, and space fields that may eventually far exceed
any benefits conceived originally. (The question of whether R&D
is a "good investment" for society deserves comment. Few would
disagree that, from a historical perspective, R&D in support of
the technological innovation process has contributed substantially
to the nation's economic growth and social well-being. The pertinent question at this time is whether it is paying off as well
as it has in the past, and whether it is paying off as well as
other investments. While precise answers to these questions are
difficult to obtain, presently available evidence all points in
the same direction: toward consistently high payoffs to society
and to the innovating firms. Rate of return estimates continue to
show that commercialization of R&D based innovations yields average returns in the 35% to 50% per annum range (3).)

Assuming that high average rates of return to investments in
innovation will continue to exist, why then do we look to government for support of R&D and particularly the commercialization
aspects of the innovation process? Why should government rather
than industry be expected to support such high yield activities?
The most persuasive arguments for the appropriateness of government investment vis-a-vis industry investment in R&D are:
whether there are insufficient incentives to attract industry
investment (assuming the investment will produce net social
benefits); and whether the R&D investment is more attractive than
other government investment opportunities. Below, some considerations or characteristics of applied R&D are discussed which
affect the appropriateness of government vs. industry support.
These characteristics are (see (2) for a similar discussion):

1. Technological uncertainty about the success
of R&D programs.

2. Market uncertainty about the saleability of
products embodying R&D results.

3. The existence of public goods or benefits
stemming from R&D which a private firm cannot capture
(inappropriability).

Technological Uncertainty. To the extent that an emerging
technology departs from tested and proven techniques and the
increases sought in technological capability rise, R&D becomes
increasingly risky. For example, there is always the prospect
that certain technological advances required to operationalize a
new idea for a product or process innovation 1) may never be made
or 2) may turn out to be excessively costly because of technical
problems.

The existence of these uncertainties or risks may cause the business or government executive to incorrectly estimate expected costs and the likelihood of technical success. Given the inability to correctly or objectively assess risk-uncertainty and hence expected costs, the risk-averse investor might be expected to undertake less R&D than otherwise for a fixed level of benefits likely to flow from a successful innovation. Thus, in the absence of government investment, there may be under-investment in areas of high technological risk--particularly where the costs of R&D are high and the resources of private companies are limited.

Market Uncertainty. Even if an idea for an innovation is proven successful in a technological sense, the private company may be unable to capitalize on it. In civil aviation, for example, the Concorde is a technological success, but its potential commercial success is currently being tested. It is costly to buy and costly to operate in comparison with alternative aircraft now in use.

Other than the obvious case where net social benefits are believed to be high and net private benefits low, there seems little reason for the government to invest in ventures on which private industry would lose money. However, it is clear that improved business and government decision making could be obtained if market uncertainties were reduced. While private firms are accustomed to dealing with market uncertainties, risk-taking propensities or perceptions of riskiness may restrain private investment below the socially optimal level. Once again, a rationale for (possibly temporary) government involvement to reduce market uncertainties can be envisioned.

Appropriability of Benefits. Another reason for government involvement in the R&D and innovation arena is the existence of noncapturable benefits stemming from innovation. That is, the developer of new technology may believe he will not be able to capture sufficient benefits, even though the net benefits are large, to make it worthwhile for him to invest in the necessary innovative activities in the first place. On these appropriability grounds, as R&D becomes more applied and developmental, the persuasiveness of the argument for government involvement usually declines; and as research becomes more basic, government involvement is viewed as increasingly appropriate.

On conceptual grounds, then, government involvement in support of innovation is most appropriate when the outcomes of R&D are uncertain in technical or market terms, appropriability problems are substantial, but a priori large net social benefits might be expected to emerge. While precise measures have not been developed for these concepts beyond a few special bases, there is some research support for their validity. As noted above, studies sponsored by the National Science Foundation tend to confirm the

assertion that average net social rates of return to private
innovation investments are as high, if not higher than average
private rates of return, thus upholding the inappropriability
thesis as a valid rationale for public support of innovation-
related R&D. Additionally, research continues to show that the
contribution of basic research to industrial innovation is substantial.   Furthermore, some studies suggest that the costs of
basic research, as a proportion of all innovation-related R&D
costs, are considerably higher than was previously expected (4).
This has prompted the assertion that governments that concentrate
public R&D funding on basic research rather than applied research
and development have chosen wisely (5).

How, then, can the tech transfer program director who aspires
to spend more of the taxpayer's money for the commercialization of
innovations argue his case? Imagine his confrontation with the
tight-fisted budget examiner who embraces these principlies with
a vengeance! Are there any practical situations where the arguments for public investments in commercialization are persuasive?
What do budget examiners really decide to do about these proposals,
and how have the decisions worked out? We turn to examine some
actual cases.

Federal R&D and Commercialization Programs

Given the basic goals of commercialization activities described at the beginning of the previous section, we can envision
a rough scale of government strategies to influence technological
change in the civilian economy, ranked by extent of Federal
involvement (see (6) for a more detailed ranking):

>   conduct and/or support of R&D
>   information dissemination (including "spinoff")
>   demonstrations
>   subsidies (e.g., tax writeoffs to suppliers;
>   loan guarantees to buyers)
>   direct purchases

Commercialization is not itself a type of intervention, but
rather a tailor-made strategy intended to stimulate technological
change in particular cases. Commercialization strategies may
involve a mix of any of the above. Policy assessments of commercialization thus ask: who will perform which tasks, and how will
the costs required to perform them be allocated between government
and private industry?

The Office of Management & Budget (OMB) has identified three
major classes of Federal R&D programs, each of which is augmented
to some extent by technology transfer and commercialization
activities (7). The OMB R&D categories and typical technology
transfer activities are presented below, with some comments about
inherent policy research issues and problems.

R&D in Support of General Economic and Social Needs. The
Federal Government assumes major responsibility because of a widespread belief that the private sector lacks sufficient incentives
to invest adequately in the national interest. Examples include
basic research to increase fundamental scientific knowledge, but
also encompass a full range of knowledge production and utilization activities in such fields as education, health care and
agriculture. Technology transfer activities in these areas vary
substantially in form and function, to the point of being almost
paradoxical. Thus, for example, the Cooperative Extension Service
of the U.S. Department of Agriculture is the most heavily sub-
sidized (over $500 million/year) and most comprehensive government
system for coupling research to utilization that we know about,
while at the same time being closely linked to the commercially
well-developed farm implement, seed, fertilizer and food processing industries (8). By comparison, areas such as education, with
a limited commercial infrastructure, receive much more limited
government support for research utilization and technology transfer (9, 10).
    The presumed rationale for Federal involvement in the commer-
cialization of technologies arising from these R&D activities
includes high net social benefits and the existence of market
"imperfections" that make the costs to private firms of reducing
market uncertainty prohibitively high; and/or that expected
returns are inappropriable. Examples of specific interventions to
assist in technology commercialization include the government loan
guarantee program in support of the sale of Lockheed L1011 aircraft several years ago (2). In extreme cases, such as socialized
medicine, government activities may essentially become substitutes
for private market mechanisms.
    Policy assessment problems in this category typically involve
issues of whether ongoing programs should be supported at historical levels (e.g., should the USDA's venerable Cooperative Extension Program continue to be publicly funded at the $550 million/
year level?); whether past government initiatives have resulted in
the desired effects (e.g., a presidential commission recently
studied the question of whether biomedical innovations have been
translated into practice more rapidly since the onset of increased
government support for biomedical research (11)); and whether new
programs should be launched in response to emerging problems
(e.g., should the government launch technology development programs
to assist domestic industries that are losing ground to foreign
competition?).
    Subtle problems exist in interpreting policy research results
in this area, since the rationale for government support may
change dramatically as industries evolve and mature; and as shifts
in public values and opinions affect cost/benefit calculations.
These problems become less difficult to the extent that longitu-

dinal studies are carried out to reveal evolutionary changes in success rates and in market conditions; and to the extent that cost/benefit assessments can be linked to relatively stable value systems.

R&D in Support of Specific National Needs. The government seeks to accelerate and augment the R&D of the private sector to assure or increase the technological options available to the nation during a particular time period. Clearly, the most prominent contemporary example is energy. In the mid 1970's, an important strategic weapon in energy technology transfer was believed to be the "technology demonstration program." However, due largely to cutbacks in government support for demonstration programs, energy research and civilian R&D in general have grown less than space and defense-oriented research in the last two years. OMB explains:

> This slowdown results from a number of considerations including, for example:
> - the need to avoid overtaking activities that are more appropriately those of the private sector such as developing, producing, and marketing new products and processes, as in the case of solar heating where the need for additional Federal demonstrations is diminished by the rapid growth of private industry efforts and the incentive provided through tax credits for increased private investments;
> - the need to avoid investing in technology where user demand or future economic viability and institutional acceptance is highly unlikely, as in the case of the Clinch River Liquid Metal Fast Breeder reactor demonstration (which is recommended for termination); and,
> - the need to avoid overinvesting in multiple demonstrations of somewhat similar technologies, or technologies that promise only marginal improvements, as in the case of coal gasification demonstrations.
> In short, the 1979 budget as it affects Federal investments in "civilian" R&D, where the Government is not the ultimate user, reflects a growing realization that the appropriate role of the Government is to emphasize longer-term (relatively lower cost) research for the future and new technology options rather than major commercial scale (and relatively higher cost) demonstrations. (7, p. 307)

Interestingly, while demonstrations are perhaps on the decline as technology transfer mechanisms in the energy field, the President's FY '79 budget requested $25 million for an "energy

extension service" in support of energy conservation R&D.   This is
an increase from $8 million in both 1977 and 1978.   Whether this
small investment will grow into another   (USDA) Cooperative Extension Service remains to be seen, of course.

The rationale for public involvement in technology transfer/
commercialization activities in this category is similar to the
first category, except that timing is critical.   Using the energy
situation as an example, it was taken for granted that the rate and
direction of innovation needed to be increased after the OPEC oil
embargo.   It was further assumed that existing markets and institutions would not be able to increase the range of technological
options rapidly enough to avoid incurring major cost penalties.
Thus the early conceptions of the Energy Research and Development
Administration's programs included liberal use of demonstration
plants to reduce technological and market uncertainties.   Additionally, flexible patent policies were granted the agency to
permit incentives for commercialization to be tailored to specific
technological and market conditions.

Policy assessment problems in this category are similar to
those in the first category, but with a stronger emphasis on
"stopping rules."   Since the emphasis here is on temporary intervention by government to augment an otherwise healthy and robust
market system, questions frequently arise about how and when to
terminate government involvement.   As in category one, a research
base of longitudinal studies linked to a sound conceptual model of
industrial technological change would be a valuable aid to policy
analysis and assessment.

R&D in Support of Direct Federal Needs, Such as National
Defense and Space.   The government is the sole or primary user of
the R&D results.   Civilian technology transfer programs associated
with these activities are generally labeled "spin-off" programs,
since what is being sought is a second-order or "unanticipated"
civilian benefit from R&D conducted for entirely different purposes.   The proper rationale for Federal support of such programs
is that the marginal social benefits expected should exceed marginal costs by as much or more than other comparable Federal
investments.   Whether this has been the case is open to some dis-
pute, as we shall discuss below.   There are two major "spin-off"
programs in government today: the NASA Technology Utilization
Program and the Department of Defense Scientific and Technical
Information Program.   The DOD program has been called "passive"
and the NASA program "active" by the General Accounting Office
(12).   The DOD program emphasizes information dissemination and
provides liaison personnel to facilitate linkages to civilian
agencies.   The passive nature of the DOD program is attributable
to the restriction of the DOD's R&D effort to its defense mission
and to personnel ceilings.   The NASA legislation, on the other
hand, requires the agency to seek widespread utilization of its

R&D results. Thus the NASA program includes "outreach" activities,
such as applications assistance teams and adaptive engineering
activities, in addition to information dissemination.

In spite of this permissive legislative mandate, the NASA
technology utilization program is funded at less than $10 million
per year. Several assessment studies have been launched in an
effort to document and quantify the social and private returns to
NASA's technology utilization efforts (13-17). The results of
these studies may help to determine whether the public investment
in "spin-off" programs is currently adequate. Two of the completed
studies are reviewed below.

In summary, then, there are three major categories of Federal
R&D programs that have civilian technology transfer/commercializa-
tion goals associated with them. Each has its own R&D rationale
which, in turn, influences policy judgments about the kind, extent
and duration of the government's involvement in technology commercialization. Research on the operations and outcomes of these
commercialization activities can illuminate the policy issues.
Particularly important are studies that can detect changes over
time in the operation of private market mechanisms that stimulate
commercial interest in technology. Additionally, cost/benefit
studies are a necessity if social and private rates of return are
to be used in the decision calculus. The purpose of the following
section is to show how these measurement and assessment problems
have been handled in recent assessment studies.

A Methodological Review of Selected Policy Studies

Few of the studies we reviewed accurately could be labeled
policy assessments, though most contain elements thereof or illustrate techniques or approaches that could be used in policy assessments. Policy assessments can be classified along two dimensions:
the time perspective taken (prospective vs. retrospective) and the
valuation framework employed (formal cost/benefit vs. various
measures of success). The following table illustrates the classification scheme and places the studies reviewed within that
scheme.

I. Classification of Policy Assessment Studies

	PROSPECTIVE	RETROSPECTIVE
COST/BENEFIT	SERI Mathematica "applications"	Mathematica "spinoff"
"SUCCESS"	RAND/breeder reactor	A.D. Little SRI International RAND/demonstration

We can summarize the advantages and shortcomings of each of these four types of assessments, without necessarily referring to specific studies. Prospective, formal, cost/benefit assessments are conceptually powerful, permit a certain degree of replication, and tend to present underlying assumptions directly. To the extent that they also use quantitative modeling techniques, they permit sensitivity analyses and are useful aids to thought. On the other hand, this approach can be misleadingly precise, since judgments and simplifying assumptions lie behind many of the numbers used. The concept of benefits is usually narrow, confined to reduced costs to producers and consumers so that analytic tools such as consumers' surplus can be used. Finally, it is difficult to define and measure outcomes using this approach, since for policy assessment purposes these outcomes should be conceived as full social costs and benefits to the Nation.

Prospective assessments that use success measures such as extent of diffusion or commercialization avoid the problem of determining net social benefits for alternative courses of action, but this, of course, omits the explicit valuation of the outcomes of those alternatives. Such studies are much less useful as policy assessments because they do not present findings that enable one to determine whether the government is justified (from a full social benefits perspective) in undertaking a particular action. Further, they are more difficult to replicate and, therefore, may be less credible to some audiences. For other audiences, this may be an asset: the use of historical data and analysis by analogy can lend credence because of the close tie to actual experience.

Retrospective studies employing success measures permit a high degree of replication because they employ empirically-derived data. Generalizations about the consequences of alternative government roles under different conditions are, in principle, possible, though the complexity of the phenomena involved create major research design problems. As in the case of prospective studies using success measures, they are conceptually less clear-cut than cost/benefit approaches and leave the question of net social benefits unaddressed. Retrospective studies of the cost/benefit type add the strength of empirically-derived data to their conceptual power, but suffer from the problems of numerous simplifying assumptions required.

Interestingly, none of the studies reviewed directly addressed the criteria for government intervention (or support of transfer activities) presented at the outset of this paper. The "ideal" policy assessment, then, would first develop data on the existence and extent of technological uncertainty, market uncertainty, appropriability, and "national need," and then assess alternative government actions, if any such actions are warranted, using one of the four types of methodologies we just described.

Cost/Benefit Studies. Two studies by Mathematica, Inc., are among the current best efforts to assess programs intended to create or facilitate the secondary application of Federally-sponsored R&D in the civilian economy (16, 17). The two studies are quite clearly operational assessments in that they develop data on the costs and benefits of secondary applications of NASA technology, but do not address the question of the value of alternative government roles or strategies for achieving the same objective.

The earlier study (16) sought to develop preliminary estimates of the economic benefits to the U.S. economy from secondary applications of NASA technology in general, and therefore did not focus on a specific program. However, to achieve that objective, the researchers used four historical case studies of particular technologies that had received NASA R&D support. (The four cases were cryogenic multilayer insulation, gas turbines for the generation of electric power, integrated circuits, and computer assisted structural analysis (NASTRAN).) The study's basic approach was to estimate the total economic benefits to the nation resulting from the technology in question, estimate the benefits that would have resulted if NASA had not contributed to its development, and subtract. National benefits from technological change were estimated using the consumers' surplus concept, where the savings accruing to both buyers and producers as a result of the cost reductions made possible by the new technology represent the national economic benefits of the technology. The premise is that NASA R&D led to an earlier realization of the technological changes under consideration. Though conceptually strong, this approach is weakened because data on the extent to which NASA contributed to the technological advance in question, and in particular the amount by which NASA shortened the introduction and use of the technology, were obtained by judgments from "experts." In addition, the noneconomic benefits and costs of the overall technological changes studied and of NASA's contribution were not addressed.

The second, more recent study (17) analyzed the costs and benefits of selected NASA Technology Utilization Office activities and, therefore, is also an operational assessment. Unlike the first study, however, both information dissemination and applications projects were studied and, in the case of the applications projects, the technologies had not yet reached the market. Two information projects and nine applications projects were analyzed. Again, the basic conceptual approach to benefits calculations was consumers' surplus. Of particular interest here is the technique used to arrive at estimates of cost savings, and thus benefits, when no sales or market penetration data exist. (The general procedure is discussed on pp. 103-123.) First, market size estimates were made, then estimates of the costs, performance, and market penetration of each project technology and competing

(baseline) technologies were developed. Estimates of the speed of
market penetration were based on use of the logistic curve, with
parameters based on historical data describing the rate of penetration of similar technologies. The savings or benefits due to
the NASA-sponsored technology equalled the difference between the
full cost of the baseline technology and the full cost of the
project technology multiplied by the number of units of baseline
technology that would be displaced by the project technology.
Finally, the result is multiplied by an estimate of the probability that the project innovation will ever reach the market.

Clearly, the validity of this procedure depends heavily on
the assumptions made and the accuracy of the estimates used.
Also, since the projects selected are not a random sample, the
results cannot be used to assess the benefits of the entire
applications program. From a policy assessment perspective, the
results of studies like this could be used to help determine
whether the government should have any role at all in the development of a particular technology (that is, whether the net benefits
would be positive). However, the weaknesses in these kinds of
approaches (to be discussed in greater detail below) make them
dubious aids to policymaking.

Policy assessments generally are conducted to inform a future
decision. They can rely largely on historical data derived from
situations deemed similar to the forthcoming decision, or they can
rely largely upon formal techniques for reaching estimates of the
consequences of alternative courses of Federal action.

Costello and his colleagues at the Solar Energy Research
Institute investigated the costs, benefits and risks of a proposed
8 year, $380 million program to accelerate the market and indus-
trial development of photovoltaic systems(18). The analysis
focused on the incremental costs and benefits of the proposed
program, with continued Federal R&D taken as the "base case."
First, the researchers estimated the response of the photovoltaic
supply industry to the various alternatives by using data from
three different sources: a workshop of photovoltaic industry
representatives, an assessment by an independent market research
firm with photovoltaic experience, and a joint SERI/Jet Propulsion
Laboratory analysis of the photovoltaic industry. Market estimates
were derived from reviews and comparison of several available market studies and from a workshop attended by representatives of
potential buyers in selected markets. Then, using a consumers'
surplus approach to calculate benefits, the authors used changes
in price and quantity estimates attributed to the initiative to
calculate the expected marginal net benefits of the initiative.
Because of uncertainties in the size of potential intermediate
markets, the effectiveness of the initiative was analyzed under a
range of possible market scenarios.

The approach is thus similar to that used by Mathematica (16)
to estimate the benefits of secondary application of NASA

technology; the weaknesses are similar as well. Both studies
required the use of market penetration models based on data of
varying quality and on dubious assumptions. As noted in a recent
critique of market penetration models, s-shaped (logistic) diffusion curves

> are not based on a theoretical explanation of cause
> and effect. First, the curves are constructed in
> large part from examination of how previous innova-
> tions diffused. Second, the historical curves show
> how a technology diffused through time but do not
> explain why it diffused.... Why a solar technology
> should diffuse like a collection of past innovations
> is an open question. (19)

Notwithstanding these shortcomings, the SERI study is a
systematic, quantitative effort to explore whether in a particular
case the Federal government's role should be restricted to R&D or
whether it should be expanded in an effort to accelerate the
commercialization of a technology.

"Success" Studies. The Rand Corporation recently conducted a
study to determine the conditions under which Federally funded
demonstration projects are effective instruments of government
action, and to identify those project organizational, funding,
management and dissemination factors that are associated with
successful outcomes (20). The projects selected for study were
all at least partially Federally supported, involved technology,
and included the private sector as either manufacturer of the
technology or potential adopter. Twenty-four case studies
provided data for the analysis. One contribution of the Rand
study was the development of measures of project outcome that
could be applied consistently across the 24 cases to ascertain
the extent to which each case represented a "success." (The three
types of measures were information success, application success,
and diffusion success.) Analysis then consisted of relating the
presence of particular project attributes to measures of success;
the attributes consisted of:

> technical uncertainty
> cost or risk sharing with local participants
> source of initiative for the demonstration
> existence of a strong industrial system for
>   commercialization
> inclusion of all active participants in the
>   technology delivery system
> absence of significant external time constraints.

Using the Rand conceptual and measurement approach, one can
make partial operational assessments of demonstration programs in
the sense that the project's success along three dimensions can
be determined. These success measures do not permit judgments of

the benefits of each project to be made directly, but they do
enable policymakers and program managers to learn whether particular projects have been effective in achieving their intended
goals. More broadly, the Rand study produced findings that serve
as guides for decision makers interested in knowing when (under
what conditions) demonstration programs "work," and what they
should do to maximize the likelihood of project success. Several
findings are particularly pertinent to policy assessment issues.
Where projects involved non-Federal cost sharing, significant
diffusion success ensued, but where there was no local cost
sharing, little or no diffusion resulted. Second, projects ori-
ginating from private firms or local government resulted in
greater diffusion success than did those originated by the Federal
government. Finally, if demonstration projects included potential
manufacturers, potential purchasers, regulators, and other target
audiences in their planning, they resulted in greater diffusion
success. In short, where the government has decided to support a
demonstration project, (i.e., the strategy has been chosen and a
decision made that there will be some Federal role), the government should share risks and costs, respond to local initiative,
and open the planning process to outside influence if the probability of success is to be maximized.

A forthcoming study by SRI International, while focusing on
the management of Federal R&D programs and projects intended for
commercialization, nonetheless offers some interesting information
pertinent to both operational and policy assessments (21). In
this study, statistical relationships were developed between a
variety of variables representing "input" factors (management
practices, project characteristics, market characteristics, performing organization characteristics, and external factors) and
a carefully designed measure of commercialization status:

    marketing has begun and is profitable
    marketing has begun and is not yet profitable
    marketing is planned but not yet begun
    marketing was started, but subsequently stopped
    marketing is not planned.

Data were collected via personal interviews with Federal program
and project managers, R&D performers, agency R&D directors, and,
in some cases, potential manufacturers for 46 R&D projects from
11 Federal agencies. The 46 projects, all of which were technical
successes, were randomly selected from two equal groups, one of
which included commercially successful projects and the other
unsuccessful ones. While it is not feasible or appropriate to
review here the findings of the study, the design of the study and
the measures used for commercialization outcomes could be used for
both operational and policy assessments. In the latter case,
analysis of projects that exhibit variations in Federal cost and/or
risk sharing would yield conclusions about the degree of Federal
involvement that leads to successful commercialization. The major

drawback of the SRI study is its heavy reliance on interview data, which proved to have low internal validity and reliability. (For example, interrespondent agreement on the same question in different questionnaires was low.)

One recent study conducted by Arthur D. Little, Inc., sought to "better understand how Federal funding of civilian research and development has functioned as an agent of technological change in the private sector." (22) Our interest in the state-of-the-art in policy assessment leads us to focus on three tasks posed in the study: to identify alternatives to R&D funding which can achieve the same objectives, to assess the "efficiency" of Federal R&D relative to these alternatives in achieving the stated objectives, and to assess the efficiency of existing Federal policies toward the support of civilian R&D. As stated, the first two would appear to be policy assessments and the last a kind of operational assessment aggregated over all Federal civilian R&D programs. To respond to these tasks, the researchers conducted six case studies of Federal involvement in civilian R&D covering four broad economic sectors: energy development, environment, transportation, and agribusiness. The sectors and specific programs selected for the cases were intended to meet a large number of criteria, including a mix of Federal funding policies, objectives, levels of effort, institutional characteristics, variations in R&D intensity, industry size and age, market structure, risk environment, and public policy regimes.

Using printed materials and interviews, in each case the researchers described and judged the outcome of the R&D program, identified the factors that appeared to have affected that outcome, and judged both the effect of R&D relative to other public policy influences, as well as the effect of hypothesized changes in the Federal role. The large number of variables involved, the complexity of the phenomena, the differing goals of the six programs studied, the lack of consistent measures of R&D program output, and (thus) the varying "success" criteria used by the researchers meant that only very speculative conclusions could be drawn. One such conclusion was that R&D alone is an ineffective influence on technological change in the private sector and that (by implication) R&D must be accompanied by other policies and actions such as subsidies and regulation if civilian sector technological change is to be influenced significantly. While useful insights can be gained from reading the individual case studies, the A.D. Little study could not guide policymakers deciding in particular situations what the government role should be or how the costs of R&D should be divided between government and industry. Use of the word "efficiency" of R&D programs suggests that operational assessments are being made, yet lack of a consistent measure of the outcome of R&D programs that could be applied across the six case studies precludes reaching conclusions about whether the benefits of the programs studied, either individually or collectively, were worth the cost.

Federal planning for commercialization is the focus of
another major study that analyzes policy options before-the-fact.
Johnson, et al. at the Rand Corporation assessed nine alternative
institutional arrangements for developing and commercializing the
breeder reactor, ranging from those with heavy private sector
initiative to complete government ownership and control (23).
Each of the nine alternatives was evaluated by the following
criteria:

> degree of clearly defined, centralized management
>   control
> effectiveness of cost control
> strength of vendor-utility interface
> value of information produced for subsequent
>   commercialization
> ease of financing
> prospects of risk-sharing
> overall plausibility.

Data and information for each alternative were developed from
past research findings, evidence from past history, and cost/
benefit studies of breeder development; considerable reliance was
placed on experience with industry-utility-government relationships during the era of light water reactor development. Evaluation of alternatives was based on a synthesis of existing, similar
cases (e.g., Dresden and Shippingport reactors; TVA and Comsat
institutional arrangements), cost estimates and levels of uncertainty based on cost/benefit studies of the LMFBR, and the history
of incentive contracting results. The strengths of the study are
its systematic structure, particularly the evaluation criteria for
institutional alternatives, and the synthesis of a variety of types
of analytic approaches in order to minimize the effects of the
weaknesses of each. Among all the studies we have considered, it
is probably the clearest example of a full policy assessment of
commercialization alternatives.

Conclusions

The policy research studies reviewed above provide interesting
examples of emerging approaches to the assessment of science and
technology policy issues through empirical research. Studies of
technology transfer processes and mechanisms are not new, but
efforts to provide a unified policy framework within which their
contribution to social and economic well-being can be understood
and assessed are just beginning to surface. Policy decisions over
the last two years regarding public funding of technology demonstration programs suggest that Federal policymakers believe there
are better investments, in terms of net social returns, than technology demonstrations. Whether these judgments will be born out
across fields of technology and areas of Federal mission

responsibility remains to be seen. As research results emerge from empirical studies of technology transfer programs and conceptual advances are made in our understanding of the dynamics of market adjustments associated with technological change, we can expect more forceful and persuasive policy assessments to evolve. (A case in point is a recently published re-analysis of the technology demonstration programs studied by Rand Corporation in light of emerging understanding of the relative strength of market incentives vs. government interventions. The analysis by Abernathy (6) suggests the hypothesis that demonstration projects are most successful when the public and private incentive mechanisms are complementary and the technological advances are incremental.)

The magnitude of the policy research challenge that faces us can be illustrated by reflecting on the very real likelihood that proposals will be put forward in government very soon to: (e.g.) increase funding for energy technology extension systems; institute a cooperative extension service for small business; implement a cooperative technology program for industries troubled by international competition; and/or fund a full scale demonstration of the next major energy technology. At this point, we have a few ideas about how to study the problems, but we really don't know the answers. In the meantime, as good Bayesian analysts, we will be able to exercise our intellects by pondering whether the social value of the $550 million/year Cooperative Extension Service, the $9 million/year NASA Technology Utilization program, and the $25 million/year energy extension service are equal at the margin.

Note: The views expressed in this paper are those of the authors, and do not necessarily reflect the official positions of the National Science Foundation, or the Solar Energy Research Institute.

LITERATURE CITED

1.   Rettig, Richard A., James D. Sorg, and H. Brinton Milward, "Criteria for the Allocation of Resources to Research and Development: A Review of the Literature", Final Report to the National Science Foundation, Ohio State University, Columbus, Ohio, 1974.

2.   Stever, H. Guyford, Dennis Shiffel, and Alden S. Bean, "Future of Aviation: Considerations Affecting the Government Role in Aeronautical R&D", in The Future of Aviation, Volume II, U.S. Congress, House Committee on Science and Technology, 94th Congress, 2nd Session, July, 1976.

3.   Mansfield, Edwin, John Rapoport, Anthony Romeo, and George Beardsley, "Studies of Social and Private Rates of Return from Industrial Innovations", Final Report to the National Science Foundation, University of Pennsylvania, Philadelphia, Pa., September, 1975.

4. "Technological Innovation: Its Environment and Management,"
    (U.S. Department of Commerce, Washington, D.C., 1967).

5. Stead, H., "The Costs of Technological Innovation", Research
    Policy, Volume 5, Number 1, January, 1976.

6. Abernathy, William J., and Balaje S. Charavarthy, "Govern-
    ment Intervention and Innovation in Industry: A Policy
    Framework", Working Paper #HBS 78-4, Division of Research,
    Graduate School of Business Administration, Harvard Univer-
    sity, Boston, Mass., February, 1978.

7. "Special Analysis P Research and Development", Special Analy-
    ses, Budget of the United States Government, 1979, Office
    of Management and Budget, Washington, D.C., January, 1978.

8. Rogers, Everett M., J.D. Eveland, and Alden S. Bean, "Extend-
    ing the Agricultural Extension Model", Final Report to the
    National Science Foundation, Institute for Communication
    Research, Stanford University, Palo Alto, Ca., September,
    1976.

9. 1976 Databook: The Status of Education Research and
    Development in the United States, National Institute of
    Education, U.S. Department of Health, Education, and Welfare,
    Washington, D.C., 1976.

10. Bean, Alden S., and Everett M. Rogers, "Staging and Phasing
    Issues in the Development of a Dissemination/Feedforward
    System in Education", in Information Dissemination and
    Exchange for Educational Innovations, Michael Radnor,
    Durward Hofler, and Robert Rich (eds.), Center for the
    Interdisciplinary Study of Science and Technology, North-
    western University, Evanston, Ill., December, 1977.

11. U.S. President's Biomedical Research Panel. "Report of the
    President's Biomedical Research Panel, Supplement 1. Analy-
    sis of Selected Biomedical Research Programs: Case
    Histories." Washington, D.C.: U.S. Department of Health,
    Education, and Welfare, April 30, 1976.

12. Comptroller General of the United States, "Means for Increas-
    ing the Use of Defense Technology for Urgent Public Prob-
    lems", Report to the Congress, U.S. General Accounting
    Office, December, 1972.

13. Bauer, Raymond A. Second Order Consequences: A Methodolog-
    ical Essay on the Impact of Technology, The M.I.T. Press,
    Cambridge, Mass., 1969.

14. Ginzberg, Eli, James W. Kuhn, Jerome Schnee, and Boris
     Yavitz, <u>Economic Impact of Large Public Programs: The
     NASA Experience</u>, Olympus Publishing Company, Salt Lake
     City, Utah, 1976.

15. Evans, Michael K., "The Economic Impact of NASA R&D
     Spending", <u>Final Report to the National Aeronautics and
     Space Administration</u>, Chase Econometric Associates, Inc.,
     Bala Cynwyd, Pa., April, 1976.

16. Mathematica, Inc., "Quantifying the Benefits to the National
     Economy from Secondary Applications of NASA Technology", a
     report to NASA. Princeton, N.J., March, 1976.

17. Anderson, Jr., Robert J.; William N. Lanen, and Carson E.
     Agnew, "A Cost-Benefit Analysis of Selected Technology
     Utilization Office Programs", a report to NASA. Mathtech,
     Inc., Princeton, N.J., November 7, 1977.

18. Costello, Dennis, et. al., "Photovoltaic Venture Analysis,
     Final Report", Vol. I, July 1978, SERI/TR-52-040.

19. Schiffel, Dennis, et. al., "The Market Penetration of Solar
     Energy: A Model Review", Workshop Summary, Jan. 1978, Solar
     Energy Research Institute, SERI-16, p. 57.

20. Baer, Walter S., Leland L. Johnson, and Edward W. Merrow,
     "Analysis of Federally Funded Demonstration Projects:
     Final Report", The Rand Corporation, R-1926-DOC, April,
     1976.

21. McEachron, Norman B., et. al., "Management of Federal R&D
     for Non-Federal Applications," SRI International, forth-
     coming.

22. Arthur D. Little, Inc., "Federal Funding of Civilian
     Research and Development", Vol. I: <u>Summary</u>, and Vol. 2:
     <u>Case Studies</u>. Prepared for the U.S. Department of Commerce.
     (Washington, D.C.) February, 1976.

23. Johnson, Leland, et. al., "Alternative Institutional
     Arrangements for Developing and Commercializing Breeder
     Reactor Technology", The Rand Corporation, R-2069-NSF,
     November, 1976.

RECEIVED March 14, 1979.

CASES

Federal R&D as an Internal Push for Commercialization of Technology

CLYDE McKINLEY

Air Products and Chemicals, Inc., Corporate Research Services Dept., Trexlertown, PA 18105

Introduction

The technology of liquid hydrogen production has been advanced very substantially during the past 25 years in response to government funding. Four agencies, the Air Force, the Atomic Energy Commission, the National Advisory Committee for Aeronautics and the National Aeronautics and Space Administration and several major programs, including the USAF/NACA high altitude turbojet engine aircraft, Nerva, Apollo, Skylab, Viking, Mariner Jupiter-Saturn, and now the Space Shuttle have furthered this technology growth. This technology has made available to the industrial user high purity, lower cost hydrogen as an alternate to cylinder/truck-delivered or on-site generated hydrogen.

Hydrogen Market - Today

Hydrogen is used primarily in petroleum processing and in the production of methanol and ammonia. Such hydrogen is generally produced in an on-site captive plant and is not considered part of the commercial market. Such captive plant hydrogen is not fully reported in U.S. Department of Commerce (DOC) statistics. The Industrial Gas (DOC) statistics lists Total Shipments hydrogen in the so-called merchant market. Merchant hydrogen is delivered for on-site storage and used by a wide variety of industries. Chemical processing is a big consumer of this commercial hydrogen in hydrogenation steps in the

0-8412-0507-8/79/47-105-151$05.00/0
© 1979 American Chemical Society

production of pharmaceuticals, plastics, pesticides and inter-
mediates. Reducing atmospheres containing hydrogen are important
to the metallurgical industry and in manufacture of electronic
equipment.

Hydrogenation of fats and oils in the food industry, use of
hydrogen in glass manufacture, and in cutting and welding further
illustrate the broad spectrum of applications.

Unique Properties of Liquid Hydrogen

All of the foregoing commercial applications have in common
that they use hydrogen in its gaseous form; liquid hydrogen as
such is not needed. Liquid hydrogen was needed for each of the
government programs in Table I. Hence the development of the
necessary purification and liquefaction technology. Although
hydrogen is not used commercially in the liquid form, liquid
hydrogen as a source of gaseous hydrogen offers three bonuses:
1) it is ultra-pure; 2) its handling convenience allows its use
as a backup or peak-shaving supply; and 3) it has low shipping
costs.

The very high purity of hydrogen from a liquid source arises
from the fact that at the normal boiling point of liquid hydrogen, all materials (except helium) are frozen solid, have very
low vapor pressures, and are essentially insoluble in the liquid
hydrogen. Liquid hydrogen, therefore, when vaporized, is excep-
tionally pure if no recontamination has occurred.

Liquid Hydrogen Production Capacity

Liquid hydrogen has been essential to each of the government
programs already noted in Table I. American production capacity
which has been brought into being in support of those programs is
chronologically presented in Table II. A cross section of the
American cryogenic industry is represented by the companies
involved in the development of this liquid hydrogen production
capacity. Air Products was directly involved in the design, construction, and operation of many of these plants. Some technical
aspects of the Air Products plants will be described later.

In 1952, a National Bureau of Standards managed and Atomic
Energy Commission sponsored plant of about 1,100 pounds per day
capacity was put into operation in Boulder, Colorado.

In the period 1957 through 1960 the (code name) Bear plants,
1500, 7000, and 60,000 pounds per day, and two other plants of
3000 and 16,000 pounds capacity were brought on-stream in support
of Air Force and National Advisory Committee for Aeronautics
programs, which include a high altitude aircraft powered with a
hydrogen fueled turbojet engine and liquid hydrogen motor development which helped move the United States into the Space Age.
Hydrogen is a very attractive fuel for such applications because
of its combustion characteristics and its very high fuel value
per unit weight.

Table I.

Federal Programs

° High Altitude Turbojet Engine Aircraft

° Nerva

° Apollo

° Skylab

° Viking

° Mariner Jupiter-Saturn

° Space Shuttle

Table II.

Liquid Hydrogen Production Capacity in U.S.

Date On-Stream	Permanently Shut Down	Still Operating	Capacity #/da	Location
1952	1959?	No	1,100	NBS Boulder, CO
1957 (July)	1963	No	1,500	APCI Baby Bear Painesville, OH
1957	1959?	No	3,000	Stearns-Rogers Bakersfield, CA
1957 (Dec.)	1959	No	7,000	APCI Mama Bear West Palm Beach, FL
1959 (Feb.)	1966	No	60,000	APCI Papa Bear West Palm Beach, FL
1960 (June)	1965 (Feb.)	No	16,000	Linde Torrance, CA
1962 (June)		Yes	60,000	Linde Ontario, CA
1963 (Feb.)		Yes	65,000	APCI Long Beach, CA
1963		?	2,000	National Cylinder Gas, Inc. Chicago, IL
1964		Yes	12,000	Airco Pedricktown, NJ
1964 (Feb.)	1970 (June)	No	120,000	Linde Sacramento, CA
1966		Yes	60,000	APCI New Orleans, LA
1972		Yes	12,000	Linde Ashtabula, OH
1977 (late)		Yes	60,000	APCI New Orleans, LA
1978 (late)		Yes	34,000	Linde 10- 17 T/D Ashtabula, OH

The Centaur and Saturn programs resulted in the series of plants through 1966. The Space Shuttle and commercial considerations account for the last three entries in the table.

Liquid hydrogen production has been and is essential to the government programs which require hydrogen in its liquid form. Its availability has been a significant bonus to industries requiring ultra-pure hydrogen.

Liquid Hydrogen, A Small Fraction of Total Hydrogen Production

Total hydrogen production is not reported as such to the U.S. Department of Commerce. The largest quantities are used on-site for methanol and ammonia production and in petroleum refining. Liquid hydrogen production is therefore relatively small in comparison. In 1976, the hydrogen consumed in ammonia production alone was 85 times the liquid hydrogen plant capacity. A comparison of liquid hydrogen plant capacity with U.S. Department of Commerce Total Shipment statistics[1] is made in Table III. However, the use of liquid hydrogen as a means for transporting and delivering hydrogen plays a critical role in making hydrogen available to the smaller industrial users.

Liquid Hydrogen, The Technology Growth

"The main contributions to liquid hydrogen processing development, from small liquefiers to large tonnage plants have been:
1. Increased cycle and equipment simplicity and efficiency.
2. Expander developments (cryogenic).
3. Catalyst developments (ortho/para interconversion).
4. More efficient and simpler purification systems.
5. Integration of liquid hydrogen production with a complex of various related products.
6. Hydrogen feed supply as a by-product from an existing source.
7. Improved methods of hydrogen gas generation.
Air Products, along with several other companies, participated in government R&D and mission oriented programs in which technology in the above areas was refined.

Increased cycle and equipment simplicity and efficiency have resulted through developments in machinery, cryogenic equipment, and cold box enclosures. Continuing developments concerned with the design and operation of reciprocating and centrifugal expanders have contributed to this increased simplicity and efficiency. The development of improved ortho-to-para catalyst as well as more efficient and simpler hydrogen purification systems have contributed to more efficient tonnage liquid hydrogen plants. Application of all these factors has resulted in decreased liquid hydrogen costs."[2]

Table III.

Liquid Hydrogen Capacity Compared with Hydrogen (Merchant) Shipments

| Year | Liquid Hydrogen Capacity | | Hydrogen Shipments (Ref. 1) |
	Pounds/Day	Millions Cu. Ft/Year*	Millions Cu. Ft/Year
1952	1,100	69	8,533
1957	12,600	786	11,441
1962	137,500	8,576	9,284
1966	319,000	19,896	23,359
1970	199,000	12,412	20,940
1974	211,000	13,160	29,327
1978	293,000	18,274	31,961**

*Based on 330 days/year and 378 cu. ft per pound mole

**1976 statistics given (last year available)

The above technical contributions were made possible by the growth in governmental agency requirements for liquid hydrogen. Commercial requirements would not have provided the incentive for the growth which has occurred in the past 25 years.

These areas of technology growth are illustrated in Table IV by a series of liquid hydrogen production facilities built by Air Products and Chemicals, Inc. to serve federal needs over 2 decades, 1957-1977.

The Painesville, Ohio plant went on-stream in July 1957 purifying and liquefying 1500 pounds per day of hydrogen from electrolytic cells producing chlorine. Refrigeration was by Joule-Thomson expansion, aided by precooling the feed stream in a liquid nitrogen bath containing silica gel for impurity adsorption. Chromic oxide on alumina ortho-to-para catalyst was employed in the product pot at liquid hydrogen temperatures, the ortho-to-para conversion all taking place at this low temperature. The conversion of liquid hydrogen to 95% para or higher is necessary for long term storage as the conversion of ortho-hydrogen to para-hydrogen releases heat sufficient to evaporate the liquid. Such conversion will occur slowly in the liquid phase.

Later that same year a larger plant was brought on-stream in Florida. Hydrogen was generated by steam reforming of propane. Refrigeration was supplied by Joule-Thomson expansion of 1500 psi hydrogen precooled by a nitrogen recycle loop. Catalytic conversion from normal to 95% para hydrogen was effected at the liquid hydrogen product temperature using chromium oxide on alumina gel.

While this 7000 pound per day plant was being built and brought on-stream a second plant for the same site was being designed. In February 1959 a 60,000 pound per day plant based on crude oil partial oxidation came on-stream, the need still being that of the USAF. In this much larger plant significant efficiency improvements had been incorporated. Liquid nitrogen precooling was followed by expansion of the 650 psi hydrogen stream through turbo expanders. Further efficiency was obtained through a more active iron gel o-p catalyst, distributed in the heat exchanger system from liquid nitrogen to liquid hydrogen temperatures. This distribution allows the heat liberated during the conversion to be removed at the highest possible temperature.

Four years later with NASA now in existence and requiring liquid hydrogen for its missions, a plant with new energy saving features was brought on-stream in Long Beach, California. 65,000 pounds per day of liquid hydrogen was produced from a refinery reformer off-gas stream. New refrigeration features of this plant were nitrogen recycle and hydrogen recycle at 1500 psi with reciprocating hydrogen expanders. A more active ortho-para conversion catalyst, was incorporated and again located at several temperature levels from liquid nitrogen to liquid hydrogen.

Table IV.

LH$_2$ Plants Show Technology Growth

Date	Location	Plant Capacity Pounds/Day	Feed Source	Federal Customer	Technology Highlights
1957(7)	Painesville, Ohio	1,500	Electrolytic Cell Hydrogen	USAF	Joule-Thomson in 1500 psi H$_2$, with LN$_2$ bath, Cr$_2$O$_3$/Al$_2$O$_3$ cat. in LH$_2$ bath
1957(12)	West Palm Beach, Florida	7,000	Steam Reforming of Propane	USAF	Joule-Thomson in 1500 psi H$_2$, nitrogen recycle, o-p catalyst chromic oxide on alumina in LH$_2$ bath
1959(2)	West Palm Beach, Florida	60,000	Crude Oil Partial Oxidation	USAF, NASA	Liquid nitrogen precooling, turbo expanders in 650 psi H$_2$, o-p catalyst iron gel and APACHI
1963	Long Beach, California	65,000	Refinery Reformer Off-Gas	NASA, AEC	Nitrogen recycle, H$_2$ recycle at 1500 psi with reciprocating expanders, o-p catalyst APACHI
1966	Michoud, Louisiana	60,000	Steam Reforming of Natural Gas	NASA	Nitrogen recycle, H$_2$ recycle at 1500 psi with reciprocating expanders, o-p catalyst APACHI
1977	Michoud, Louisiana	60,000	Steam Reforming of Natural Gas	NASA	Nitrogen recycle, H$_2$ recycle at 1500 psi with reciprocating expanders, o-p catalyst APACHI

Three years later another large plant was brought on-stream near New Orleans, Louisiana at Michoud. The hydrogen source was steam reforming of natural gas.

A decade later another plant was built at Michoud to serve the Space Shuttle program. This plant was similar to the earlier plant at Michoud, in fuel source and in process cycle.

The extensive involvement by Air Products in the liquid hydrogen program, provided the base for its exploration of and development of commercial markets best served by liquid hydrogen.

The technology improvements highlighted in Table IV would not have occurred in this two decade time frame without the federal agency market.

Safety

Great growth in understanding of the hazards and risks associated with handling large volumes of liquid hydrogen took place among the industrial-federal teams of producer and user. It became possible to produce, handle, and use liquid hydrogen with the confidence that safe, proven, and understood procedures were being used. Much excellent liquid hydrogen safety literature exists. A few major concerns which were resolved in the 50s and 60s are listed here.

A major liquid hydrogen spill was of great concern. Would such a spill result in the potential of an open atmosphere detonation? Experimental work showed that an open atmosphere detonation was extremely unlikely; very strong ignition and substantial confinement (as opposed to open atmosphere) would be required to yield a shock wave upon ignition.

Measurement of the solubility of solid oxygen in liquid hydrogen (and low temperature gaseous H_2) showed exactly what had to be done in O_2 removal during the H_2 purification process to avoid solid O_2-LH_2 explosions. Understanding of another oxidant of concern, N_2O, was also obtained. N_2O may be present in hydrogen from electrolytic cells but it can be converted catalytically in H_2 to water and N_2 which in turn are removed by conventional means.

Gaseous hydrogen, containing a suspended second phase was found (as expected) to generate static (promote charge separation). This added to understanding of the need to avoid static inside purification systems where condensed oxidant phases could be in contact with H_2.

The growing technology provided experience in coping with the more conventional cryogenic hazards associated with material's brittleness, with cold flesh "burns," and with liquid to gas expansion in confined spaces.

Because of the federal funding the hydrogen technology was quite well publicized, and especially all of that which related to safety.

Conclusions

The technology of liquid hydrogen production was significantly developed, refined, and practiced as a result of federal needs and the associated federal funding. Such development would not have occurred in the same time frame with only commercial markets as the incentive.

Commercial applications have been able to take advantage of the three liquid hydrogen bonuses: 1) handling convenience, 2) ultra purity of gaseous hydrogen obtained by evaporation of the liquid, and 3) low shipping costs.

References

1. U.S. Department of Commerce, Current Industrial Reports, Industrial Gases.

2. Newton, Charles L. (Air Products and Chemicals, Inc.) Hydrogen Production, Liquefaction and Use (two installments) August and September 1967, Cryogenic Engineering News.

RECEIVED March 14, 1979.

Commercialization of a New Starch-Based Polymer

WILLIAM M. DOANE

Northern Regional Research Center, Agricultural Research, Science and Education Administration, U.S. Department of Agriculture, Peoria, IL 61604

This paper provides an example of commercialization of a product invented during in-house federal research. Various events are described that played a role in transferring the technology to the private sector.

The product was a starch-based polymer with unique properties for absorbing large amounts of aqueous fluids. The research leading to the absorbent product is part of a program directed towards developing renewable agricultural commodities as partial or total replacements for petroleum-derived products. In this specific research project, we have been studying the chemical bonding of synthetic polymers to starch, a natural polymer produced in great abundance in many agricultural crops.

Our early research pointed out that the best way to covalently bond synthetic and natural polymers was via a method referred to as graft polymerization. In this technique, reactive sites are formed on the starch backbone and then the appropriate monomer (the individual building unit of the polymer) is brought into contact at the reactive sites and caused to polymerize. Acrylonitrile, a polymerizable monomer, readily graft polymerizes onto starch to yield a copolymer in which the synthetic polymer, polyacrylonitrile, is covalently bonded to starch. Treatment of the starch-polyacrylonitrile (S-PAN) with sodium hydroxide converts the S-PAN to a highly hydrophilic composition possessing excellent properties for a thickening agent.

Although the thickening properties of the hydroxide-treated S-PAN were predicted and, in fact, were the properties being sought, an unexpected property of the polymer, that of water absorbency, was not expected. We found that on drying the thickened dispersion, a solid product was obtained which, when added to water, would absorb hundreds of times its weight of water but would not redissolve. The initial observation of this property was made when a film that formed on evaporation of a thickened dispersion of hydrolyzed S-PAN was placed in a shallow tray containing water. The film rapidly imbibed the water and increased in surface area about thirtyfold. The swollen film

showed an increase in weight of about 300 times over the dry
film. (Further studies of this polymer resulted in products that
would absorb 2000 times their weight in water.)

Discussions among the group involved with the discovery of
the absorbent and a search of the literature to help assess its
uniquness resulted in a somewhat different approach to reporting
the discovery than was usual in our Laboratory. Compiling a list
of potential applications, where enhanced absorption of aqueous
fluid would be desirable, caused us to report the discovery not
only in the scientific literature but also in trade journals and
the popular press.

Little did we realize at the time the impact our information
officer, who was responsible for preparing news releases for the
popular press and trade journals and magazines, was to have on
the successful commercialization of the product. As the four
scientists were describing the product to him and demonstrating
how rapidly the product would absorb hundreds of times its weight
of water, he quickly gave it the name Super Slurper. He explained,
much to the chagrin of the four scientists, that a name other
than hydrolyzed starch-polyacrylonitrile graft copolymer was
needed if we were to communicate with the public. We now recognize how right he was and how significant a role the name he gave
to the absorbent has played in promoting the product.

He prepared several news releases whose contents varied
depending on the audience he intended to reach. Largely through
his efforts, our Center received several hundred inquiries in the
first few months for more information on the Super Slurper. (We
estimate that over the 5-year period since the first announcement,
we have received and responded to over 5000 inquiries.)

It had been decided before the first news release was sent
out that we should prepare some printed material in addition to
the scientific paper we had written. An information sheet was
prepared that was of more use in responding to the general
inquiries than was the scientific paper.

Our next decision, that of providing small samples of the
absorbent, played, I believe, a very significant role in the road
to commercialization. We realized that the small sample (a few
grams) was insufficient for evaluation in an end-use application,
but it did serve to further pique the interest of the recipient.
Continuing requests for samples caused us to turn away from
laboratory glassware and to a larger reactor in which a few
pounds of the starch product could be prepared. It should be
mentioned here that our mission is to conduct basic, long-range
research of a high-risk nature that the private sector does not
carry out. We do not perform the development research which the
private sector, with its expertise, can do so much more efficiently.
hus, in going to a larger reactor, engineering or development
studies were not undertaken.

Preparation of the larger quantities did serve to demonstrate
the feasibility of making the polymer in systems other than

laboratory glassware. It also provided us with enough information
to enable us to come up with a preliminary cost-to-make estimate.
This rough estimate permitted us to respond to the question on
cost of the polymer that came up so often. The paper we wrote
covering the larger scale preparation and the cost estimate
turned out to be quite useful, especially to the small company.

   We used the great interest shown in the polymer and some of
the feedback from those receiving samples to attempt to encourage
private industry to undertake development studies on Super Slurper.
The discovery was patented and royalty free, non-exclusive licenses
were available from the U.S. Department of Agriculture. The
reluctance by private industry to undertake development of Super
Slurper without a proprietary position was partially overcome,
when it was recognized that innovations arising during design of
a commercial process might well offer them the opportunity for
patenting. Another incentive for the industry was that we would
provide the names of those licensees who were producing the
product in developmental quantities to all who contacted us about
Super Slurper. In order to do this, we required a letter from
the licensee stating that they would respond to all inquiries
they received. As of this writing, the U.S. Department of Agriculture has issued 42 licenses, and 5 of the licensees have asked
to be listed as suppliers of developmental quantities. We have
been told by one of the suppliers that our listing of their name
had resulted in over 1000 inquiries.

   In late spring of 1978, the first company to obtain a license
opened a plant and started commercial production of the absorbent
polymer. Another of the licensees has been producing several
thousand pounds per month for nearly a year. Some others have
informed us they are now completing pilot-plant studies.

   As increasing commercial quantities become available, the
list of uses for Super Slurper grows rapidly. Currently we are
aware of its use in such diverse areas as disposable soft goods
to absorb body fluids, removing water from pulverized coal, seed
and root coatings, thickening water in fighting forest fires,
hydroseeding to establish plant growth on new construction sites,
removing traces of water from organic solvents, and as an absorbent in hand powder for athletes. We have been informed by the
private sector that their market estimates suggest a U.S. market
of about 1 billion pounds per year for Super Slurper.

RECEIVED March 14, 1979.

INDEX